ROUTLEDGE LIBRARY EDITIONS: TELEVISION

Volume 12

SATELLITE BROADCASTING

SATELLITE BROADCASTING
The Politics and Implications of the New Media

Edited by
RALPH NEGRINE

Routledge
Taylor & Francis Group

LONDON AND NEW YORK

First published in 1988

This edition first published in 2013
by Routledge
4 Park Square, Milton Park, Abingdon, Oxon OX14 4RN
605 Third Avenue, New York, NY 10017

Routledge is an imprint of the Taylor & Francis Group, an informa business

British Library Cataloguing in Publication Data
A catalogue record for this book is available from the British Library

ISBN: 978-0-415-82199-5 (Set)
eISBN: 978-0-203-51517-4 (Set)
ISBN: 978-0-415-83926-6 (Volume 12) (hbk)
eISBN: 978-0-203-77345-1 (Volume 12)

Publisher's Note
The publisher has gone to great lengths to ensure the quality of this reprint but points out that some imperfections in the original copies may be apparent.

Disclaimer
The publisher has made every effort to trace copyright holders and would welcome correspondence from those they have been unable to trace.

SATELLITE BROADCASTING

Satellite technology seems likely to change the face of broadcasting both within nations and internationally. This book provides a thorough examination of the possibilities and key issues. It begins with a guide to the technical development of different systems of satellites and signal reception and an outline of the international, political and regulatory issues involved. It then examines the current situation in various industrialised countries by analysing current interest, launching plans, funding, the interaction between satellite, cable and VCRs and the effect, both current and potential, on existing broadcasting systems. The book is concerned throughout with a wide range of cultural considerations and the potential impact of the new media on existing broadcasting systems.

SATELLITE
BROADCASTING

The Politics and Implications of the New Media

Edited by
RALPH NEGRINE

ROUTLEDGE
London and New York

First published in 1988 by
Routledge
a division of Routledge, Chapman and Hall
11 New Fetter Lane, London EC4P 4EE

Published in the USA by
Routledge
a division of Routledge, Chapman and Hall, Inc.
29 West 35th Street, New York NY 10001

Printed and bound in Great Britain by Mackays of Chatham PLC, Kent

British Library Cataloguing in Publication Data

Satellite broadcasting: the politics and
    implications of the new media.
    1. Direct broadcast satellite television
    I. Negrine, Ralph
    384.55′4   HE8700.7
    ISBN 0-415-00109-9

Contents

Tables and Figures

Introduction

Satellite Broadcasting: An Overview of the Major Issues

Ralph Negrine

The last decade has seen an enormous change in the television broadcasting scene across the world. Cable systems and satellite broadcasting — and particularly the marriage of these two technologies — have brought about a phenomenal increase in the available channels of television entertainment and video communication. Broadcasting by satellite has enabled the subscriber to a cable system, whether in the USA or Britain or Sweden, to gain access to a wide range of material previously undreamt of. But these two technologies have other implications which are usually less visible: both technologies are important culturally and industrially. For these reasons, it is necessary to consider them not only within the context of more broadcasting entertainment but also within the context of national states planning for their future industrial needs and cultural desires. And, because the various forms of broadcasting by satellite straddle national frontiers, there are implications for those nations which desire to retain and encourage their own cultures in the face of foreign broadcasting.

But beyond these rather general statements about the implications of the new media, there is an underlying complexity that cannot be submerged in generalities. Each nation state will experience these changes in a different way and, even more importantly, not all nations will perceive the new media as a threat to their industrial development and to their existing structures of broadcasting. Many are likely to welcome the additional channels of entertainment, the diversity in telecommunications delivery systems and the new era in broadcasting competition. Similarly, it is possible that whilst policy makers and public service broadcasters may fear the impact of the new media, the viewing

1

public will look to them for an escape from the limited fare currently on offer.

General statements about the new media and their implications are, therefore, likely to be of limited value. Certain themes, however, do tend to repeat themselves in certain contexts. There is some sort of unity of concern across Europe, for example, where public service broadcasting organisations under some form of state control have until the recent past been dominant. These organisations are soon to face immense competition from an array of eager entrepreneurs as their monopolistic position is rapidly eroded.

If one were to look forward to the beginning of the next decade — and assuming that all current plans for new television services come to fruition — television broadcasting across Europe will have been transformed to an extent unimaginable a mere ten years earlier. Cable systems will exist side by side with 'low', 'medium' and 'high' power satellite systems as well as the traditional forms of terrestrial broadcasting systems. Each of these systems will attempt to offer something unique and the viewer will face the unenviable task of exercising a very difficult choice. What that choice will be is difficult to say since the situation is constantly changing; what is clear, however, is that the changes will range across hardware and software; they will impact on existing broadcasting structures, on sources of funding, on programme producers, on regulators and on many other groups and institutions which are connected, however tenuously, with 'the media', both old and new.

To illustrate this change, one need only look at Britain, the source of many worthy ideas about broadcasting and the provider of a much imitated model of a broadcasting structure with a public service broadcasting philosophy.[1] In 1982, Britain acquired her fourth — and an advertising funded — national broadcasting service. This completed the duopoly and created a system whereby two non-commercial services (BBC 1 and 2) competed with their commercial counterparts (ITV and Channel 4). Five years later, in 1987, these terrestrial services are broadcasting alongside a host of satellite delivered services; over ten new services, including the English language SKY, Super Channel, Arts Channel and Lifestyle are now available to both cable subscribers and to owners or renters of large satellite receiving dishes.

This menu will be increased as Direct Broadcasting by Satellite (DBS) services become available for direct-to-home reception on

Table I.1: Proposed European launches of satellites for TV use

DBS	Launch Date	Power (Watts)	Main Purpose	Capacity
TDF-1 (France)	mid-1988	230/260	DBS TV	4 channels
BSB (UK)	late 1989	120	DBS TV	5 channels
TV-SAT1 (FRG)	Oct. 1987	200/230	DBS TV	4 channels
Tele-X (Nordic countries)	late 1988	230	DBS TV	4 transponders
Olympus	1989	230	TV	2 transponders
Medium Power				
SES ASTRA	late 1988	50	TV	16 transponders
Eutelsat II	1990	50	TV	16 channels

small dishes. The French, German and British DBS services (TDF-1, TV-SAT and BSB respectively) are due to be launched in the remaining years of this decade and to be in operation by 1990/1. Some of their services comprising entertainment, news, films and the usual host of popular media fare, have already been announced and many others are still unplanned. To this list one must add the prospect of another 10 to 16 channels which will be made available on the SES ASTRA medium power satellite project in the same period (see Chapter 5 for details). Though a considerable re-alignment of services and satellites will undoubtedly take place, it is clear that there will be a very significant increase in the number of entertainment channels available across Europe. (Table I.1)

Whether these additional services enrich the menu rather than simply increase its length is, in one sense, an issue of little consequence. The changes currently taking place are at such an advanced stage of development that academic or even policy debates matter little in the rapidly changing world of the new media. The speed of technological change and the inability (or unwillingness) of governments across the world to plan (or place in a strait jacket) the direction of these developments, has given them the look of an unstoppable force. In some European countries there has been an acquiescence on the part of regulatory and governmental agencies — acquiescence since it is typified by an unwillingness to debate the wider issues of the place of the broadcast media in the cultural, social and political life of a nation state — which has allowed the technological rollercoaster to move unobstructed by political or social concerns.

Yet Europe is not alone in being 'infatuated' with the prospect of satellite broadcasting and in curtailing the much needed debate about the future of television. Reviewing the Australian scene in this volume, Reinecke observes a similar pattern of infatuation, of a desire to move ahead very rapidly and a reluctance to pull back in the face of difficulties. Furthermore, there appears to be a desire to avoid asking whether more channels actually increase 'diversity and choice'.

What are the implications of these changes in the available technologies and in the relevant regulatory frameworks? At this stage, no clear pattern has emerged. This is partly due to the fact that many of the satellite projects are still 'in the future'. It is, for example, possible that technical problems and launcher disasters will curtail many of the proposed projects. Similarly, governments may decide that publicly funded DBS ventures are too costly. The effect of such changes in fortune will be to reduce the number of 'birds' in space. However, it is difficult to foresee a situation where no satellites will be operational. In other words, it is likely that satellite broadcasting in some form — and perhaps even as a pale imitation of the promise of abundance — will exist in the near future.

Under such circumstances, television will become a much more competitive industry. With increased competition for the television viewer, it is probable that only those organisations which are able to satisfy him/her are likely to survive. This will apply to the public service organisations as well as to the commercial broadcasters despite their very different sources of revenue. The former will have to justify the continued existence of what many people see as an outmoded and regressive tax, namely the licence fee, in the face of an onslaught from 'free' commercial television. This will be particularly true of those public service broadcasting organisations which are no longer able to count on the support and commitment of their governments. The latter will have to convince the advertisers that they can deliver a sizeable audience. In both cases, broadcasting organisations will become more aware of the competitive market place within which they will increasingly have to operate and, as the established methods of funding broadcasting come under severe strain, they will also need to explore different ways by which to get the viewer to pay for what they consume. In all these ways, the traditional relationship which had previously existed between the viewer and the public service broadcasting organisation is likely to undergo changes as the method of payment for entertainment changes. Recent evidence from the USA also points to the difficulty that 'traditional' commercial networks face when confronted by the new media. The overall effect of this on the quality and content of the media is likely to be radical and significant.

Alongside the increased competition, nations will find themselves exposed to foreign media. The long-term impact of these, usually Anglo-American, channels on local cultures is difficult to gauge but it is clear that some concern is being expressed. Most European nations are now exposed to such channels, as is Australia, Canada and to a much lesser extent Japan. Even the Third World can suffer from the popularity of satellite programming spilling over into its territory. A recent report from the Carribean Publishing and Broadcasting Association expressed concern over the islands' dependence on American TV—the islands fall within the footprints of American satellites—at the expense of locally made and produced programmes and television channels.[2]

Before exploring these issues in greater detail, it is necessary to take a closer look at the new media themselves. Section Two, below, briefly outlines the technologies currently in vogue. Section Three will examine some of the forces behind these developments and it will pay particular attention to the European context where satellite broadcasting represents a radically different form of broadcasting structure. Section Four will make reference to the other countries reviewed in this book and will briefly summarise some of the major points which emerge in the text.

The aim of this book is to review the development of satellite broadcasting in different national contexts. Individual chapters explore the cultural and industrial implications of the new media and of satellite broadcasting in particular. They also explore national and commercial strategies towards satellite broadcasting and the motivations and plans of those corporate bodies which have shown an interest in running such broadcasting systems. Since satellite communications — and satellite broadcasting is one facet of this — highlight both industrial/political as well as traditional broadcasting concerns, it is imperative that they are placed in their respective commercial, political and cultural contexts.

THE NEW MEDIA — CABLE TELEVISION, SATELLITE BROADCASTING AND DBS

What, then, are these new media and how will their separate — and combined — developments change the face of television broadcasting? To delve into these questions is to explore a scene that is rapidly changing and a scene that is populated by an array of technologies characterised by a succession of words and letters picked as if at random: cable, DBS, SMATV, Ku Band, C-MAC, D2-MAC, ECS, East Beam, West Beam, polarisation... to mention but a few. Paradoxically, describing the new media is far simpler than attempting to chart their development. The history of the new media — if 'history' is a term appropriate to a field that is of so recent origin — is full of hopes and dreams, failures and disappointments; and there are no guidelines by which to distinguish which technology or indeed which project will end up in which category.

As always, it is important not to let one's imagination run wild with thoughts of an overabundance of entertainment and information as a direct consequence of the new media. There are many real obstacles to this: some are economic or political, others are technical, for example, the availability of launchers for satellites. So caution is advised when looking at the new media: caution in dealing with the proposals, caution when dealing with the promises, caution when dealing with probable launch dates and generally caution when dealing in an uncharted and still developing area. Because of its makeup, the field of the new media is a cross between informed journalism and timely social science: a worthy study is one that offers a depth of analysis which will enable the reader to comprehend the undercurrents that will shape the contours of the television broadcasting systems of the future.

Cable and satellite — a profitable union

Cable television has, up till now, been the most widely discussed of these newer methods of signal distribution and it neatly summarises the complexity of the issues described above. Before the public can have access to cable systems, they have to be constructed and their construction — their technological makeup, the speed of construction, the terms of operation, their funding and so on — very largely depends on the political and economic considerations which inform different governments not only with regard to broadcasting matters but also with regard to the role of cable systems in the future telecommunications infrastructure of that nation state. Not surprisingly, therefore, cable systems have come to occupy a central position in discussions about both communication/cultural matters and industrial planning/regeneration.[3] As the contributors to this present book will show, satellite broadcasting offers another important example of this increasing convergence between the cultural and industrial policy-making arenas.

Interest in broadcasting by satellite has increased enormously since the mid 1970s when its potential was demonstrated by such services as Home Box Office in the United States. HBO was the first company to use satellites as a means of delivering television signals to cable systems and, today, it remains one of the most profitable organisation of its type. Its decision, in 1975, to use domestic satellites for its pay-movie cable channel was unusual as there were only a few satellite receiving dishes owned and operated by cable systems. Furthermore, the whole field of satellite communications was still relatively new; the FCC's 'open skies' policy allowing private companies to launch and operate satellites and to rent their services to commercial companies had only been adopted a mere three years earlier in 1972.

The popularity of HBO boosted interest in cable systems generally. Firstly, the number of cable systems able to collect satellite distributed signals increased rapidly. According to Hollins, in 1977 there were about 500 cable system dishes but by 1982 there were just under 5000.[4] Secondly, the number of cabled households in the United States doubled to 28 million between 1978 and 1982; in 1986, that figure stood at around 40 million. Thirdly, HBO also generated a great deal of interest amongst would-be programme or channel providers. By 1985 some fifty-five national satellite video services were available to cable operators.[5]

The success of some of these newer programme channels and of cable systems generally radically altered the face of American broadcasting. Not only was the range of programming now greater than ever before but individuals could now exercise an element of choice by lending their support to their favourite programme channel. Services which were able to satisfy their viewers would then thrive in this new market dominated era of broadcasting.

The American experience did not go unnoticed in Europe. Britain, amongst others, began to re-consider her policies towards cable systems and satellite broadcasting from about 1980 onwards when attempts began to be made to come to terms with these newer technologies of signal distribution and with their increasing interconnections. The debates surrounding cable systems and satellite broadcasting in Europe generally but also in, for example, Canada and Australia all began to reflect the growing concern over the future of communications at a time of immense and rapid technological change.

But the debates in Europe were also symptomatic of a realisation of impending change and the need to make early decisions; the launch of Murdoch's Sky Channel in 1982 as a satellite delivered television service to European cable systems served as a reminder that the American experience was going to spawn European imitators.

Sky Channel began in 1982 in a small way: it used the Orbital Test Satellite (OTS) and, at first, provided a mere two hours of programming per day. In due course it switched to Eutelsat (ECS F-1) and, by 1987, it provided a full programme service catering to music fans, sports fans and fans of general entertainment programmes. Whilst in 1985 it could be seen in under 3 million homes, in 1987 it could be seen in over ten million homes right across Europe.

Like HBO, Sky Channel was beamed direct to cable systems and so to some extent its success depended on the availability of cable systems and the existence of a favourable regulatory framework. As pressure to develop cable systems increased and as cable systems developed, the prospects of services delivered to cable systems by satellite improved. The opposite was also true; with little growth in cable, the fortunes of satellite delivered services were not likely to be promising. The necessary union of these two media works to both their advantages: those in the satellite business pay a great deal of attention to forecasts of cable growth and those in the cable business pay a great deal of attention to the product of satellite channels.

How does one assess the importance of Sky Channel (and similar, though less popular, services) in the European context? Three areas of significance immediately suggest themselves. Firstly, it demonstrates that the idea of television broadcasting across Europe is a feasible one. Secondly, it seems to confirm the belief that there is a market for popular entertainment—though all the services still make financial losses—and a market that may be fickle in its loyalty to state directed or state regulated public service broadcasting systems. The Swedish and British data (Figure 3.3 and

Table 5.3) confirm this. Thirdly, it highlights the danger of satellite services broadcasting across national frontiers and by so doing making a mockery of notions of national sovereignty in broadcasting. These three points taken together highlight a problem which Europeans had never before had to confront in quite this way, namely, the foreign invasion of national air waves.

Canadians, on the other hand are very familiar with this problem. Canada has always tried to confront its neighbour, the USA, in order to bring about the development of a Canadian identity. A practical consequence of such a policy would have been the growth of an indigenous media industry. However, this situation has not always been successfully dealt with; according to Ferguson, Canada's multiple broadcasting systems

> bring a daily deluge of American programmes across what Canada's cultural nationalists have come to regret is "the longest undefended border in the world".[6]

She concludes that the

> Canadian experience of invasive broadcasting neighbours and the popularity of the products they offer is a tale with a moral... *What the evidence of the marketplace shows is that the realpolitik of broadcasting today is the interdependence between the old programming economies and the new delivery systems which know no frontiers: Canada's experience is becoming less unique.*[7]

How, then, would European nations continue to exert control over their broadcasting systems in the era of trans-border communication? Herein lay the concerns and events which have forced both the media industry and governments to confront the commercial and regulatory challenge of the new media.

Direct broadcasting by satellite (DBS)

It would, however, be incorrect to present the new media simply as a combination of satellite and cable systems, no matter how complex that combination. There is one other new medium, Direct Broadcasting by Satellite (DBS), which promises to complicate further the pattern of developments. Unlike existing

9

Table I.2: Cable penetration in selected countries (mid 1987)

Country	Households (millions)	Homes cabled '000s	Penetration %
Austria	2.76	320	11.6
Belgium	3.72	3,100	83
Canada	8.5	5,000	59
Denmark	2.2	1,150	52
Finland	1.8	270	15
France	19.5	25	0.13
Germany	25.3	2,300	9
Luxembourg	0.141	68	48.2
Netherlands	5.5	3,100	56
Norway	1.5	400	26
Sweden	3.5	240	6.9
Switzerland	2.5	1,400	56
United Kingdom	21.8	188	0.86
USA	81	44,000	54

communications satellites of the type used across Europe by programme services to deliver signals to cable systems, DBS is aimed directly at individual households so by-passing cable systems altogether. Its potential growth is seen to be in those countries where cable has so far failed to expand (Table I.2).

Detailed planning of the 12 GHz frequency band for DBS took place in Geneva in 1977 during the World Administrative Radio Conference. For a variety of reasons, WARC '77 only dealt with Europe and Africa (Region 1) and Australasia (Region 3) leaving the Americas (Region 2) to a later series of meetings.[8] (see also Chapter 1 in this volume). In its deliberations WARC attempted to deal fairly with all the countries within each region. Thus, in respect of Region 1, it laid down two major principles which, in due course, were to prove too restrictive. Firstly, it wanted to ensure that DBS services operated within national boundaries so as to minimise interference or overspill into adjoining countries. Secondly, it allocated DBS orbital spaces equally to all the European states irrespective of whether they were likely to operate such systems or not.

Subsequent proposals, formulated within this framework, confirmed the view that WARC had been too restrictive in its work and not very appreciative of the rapid technological changes which were continually taking place. For example, plans for DBS services were initially drawn up along the lines set down by WARC: high power (230 watts) satellites available to dishes of 0.9 metres. But by 1985, technological advances had ensured that the same desired effect — individual reception on small dishes — could be achieved by medium power satellites and that the dishes could be even smaller than 0.9 metres. More importantly, though, the desire to restrict services to individual countries proved the most difficult to pursue since DBS appeared to have made the transition to a pan-European enterprise. As a result British, French and German DBS are likely to transcend national boundaries.

In practice, then, the WARC requirements exacerbate the difficulties of planning the allocation of orbital space since they impose non-technical, political boundaries on technologically advanced — and advancing — communications systems.This and the prospect of DBS-like services,for example, the SES ASTRA project (see Ch.5) available via medium power satellites to small receiving dishes, has eroded the distinctions between types of satellites and has introduced an element of confusion over the precise nature of the technologies in question.

The individual householder is unlikely to be particularly concerned about the technology *per se* but merely interested in the choice which competing technologies offer. There will, in the future, be a choice between cable subscription and the purchase of a satellite dish (or neither) and between competing programme providers. In this context, any individual decision will have to take account of the availability (or otherwise) of cable and of satellite dishes; of the often overlooked costs of TVRO decoders and subscription fees which may prove to be quite considerable; of the costs of these different media and of the choices of programmes or other non-entertainment services, for example interactive services on cable systems, which they can offer.

POLICIES AND REGULATIONS: THE EUROPEAN DIMENSION

The growth in popularity of (high and low power) satellites for the delivery of television signals and their increasing technological sophistication and power, have brought a number of policy problems in their wake. Should satellite broadcasting be used as a cultural and educational tool rather than as simply another source of television entertainment? How should a nation deal with unwanted satellite signals spilling over into its territory? And, in another but related context, should cable systems be created as mere relayers of satellite services? Should they be constructed in such a way as to become the infrastructure for all forms of communications? These questions became the more urgent as the drive to exploit the available technologies accelerated.

Individual nation states are now forced to consider not only the established structure of their broadcasting systems but also the construction and integration of the newer systems of signal distribution into their industrial and cultural sectors. The new media have long ceased to be matters of entertainment only since they have important national cultural and industrial significance.

With cable systems each nation state had proposed and implemented its own particular solution. One finds a degree of diversity that is appropriate to technologies which have, in the first instance, national impact only. But unlike cable, satellite communications is a national concern only in those few instances where the geographical spread of the nation state enables it to work freely within national boundaries. The Canadian and Indian experience with satellite broadcasting illustrate the successful use of domestic satellites for national communication.[9] But, as already noted, the Canadian experience also highlights the problems of signals being available from a powerful next door neighbour; signals which are seen to wreak havoc with the Canadian broadcasting ideal.[10]

Nevertheless, even geographically large nation states do face difficulties for beyond a certain point extra demand for satellite orbital space impinges on international considerations. The example of the USA is instructive. International telecommunications is regulated by the International Telecommunications Union (ITU) which attempts to operate rules of fairness and impartiality as between the demands and needs of all its 155 (or so) member nations. However, increased pressure for more orbital slots for American satellites can only be met by the ITU decreasing its notional allocation of spaces for other, less economically powerful, nations. So the ITU has to determine whether those who are currently able and willing to operate satellite systems (i.e. the USA) should be allowed to do so without restrictions or should such states bow down to some international agreement as to the fairest way to share out the limited orbital spaces.

12

Only about one-fifth (76 degrees) of the earth's geostationary orbit can serve the needs of the USA for continuous and uninterrupted communications. Outside this segment, there are problems in reaching either end of the nation. Yet the narrow arc presents its own problems: satellites too close together may interfere with each other's signals and though technical sophistication — narrower beams, more powerful satellites — may overcome some of these difficulties others remain. One FCC staffer is quoted admitting the enormity of these problems and the need

> to balance the conflicting desires of the applicants, the actual traffic volume and the distribution requirements of the various domsat [domestic satellites] systems, the constraints on satellite location imposed by satellite design, the announced plans of other countries, and consideration for fair treatment of existing and new domsat operators.[11]

Inevitably, this raises a range of difficult problems; problems which now have their European counterpart with the debates surrounding the proposed privately funded SES ASTRA satellite in competition with ITU Intelsat/Eutelsat satellites. Which international organisation will regulate the skies and ensure fairness and equality? Should nations that can exploit satellite ventures do so irrespective of whether they use other nation's (unused) orbital slots? Who will control the content of satellite services? How does one ensure that one country is not inundated with signals from another, possibly unfriendly, country? As many of the chapters in this book illustrate, it is becoming increasingly difficult for a nation's cultural boundaries to remain intact. It may still be possible to exercise regulatory control over cable systems (see, for example Chs. 2 and 5) but once satellite receiving dishes for DBS or satellite TV reception become more commonplace, there will be no feasible way of monitoring or controlling the public's choice of medium.

The development of these satellite broadcasting systems — whether DBS or of less powerful satellites for purposes of linking up cable systems or master antenna services (SMATV) — will depend on such issues as government policies, international regulation, the existence of cable systems, the power of satellites, the buoyancy of the advertising industry and the organisation, popularity and funding of the programme services that satellite channels will carry.

13

Across Europe, there is as yet no unity of approach to the new media, whether of cable or satellite broadcasting. Europe remains a collection of distinct countries each pursuing policies geared to their own national needs and national objectives. For example, a country such as Sweden with strict rules about advertising on terrestrial services has to ensure that foreign services beamed into its territory do not break those rules. A country proud of its culture, such as France, will be concerned about the importation of foreign programmes. A country proud of its broadcasting structures, such as Britain, would be concerned lest the ecology of its services are upset by a foreign body.

Generally speaking, then, countries which treat television broadcasting as a vital cultural and social force within their territories are confronted and threatened by services which are motivated by an overriding concern with entertainment and profit. European governments are now being pressed to legislate for the new media. The mere availability of the technologies of cable, DBS and the like, has created pressure for actions and decisions. Where governments have been able to exercise control over the direction of these developments, e.g. DBS and cable, they have developed policies which exploit their possible contributions to national objectives. In Germany, Britain and France the regulation of cable systems has been informed by a desire to use such systems as telecommunications networks rather than simply as television relay networks. The same degree of emphasis on industrial objectives can be found in the development of DBS in Germany and France, as well as in Japan and Sweden. Britain also originally embarked on a DBS policy which would have exploited indigenous advanced technology but it found this policy reversed as costs became too prohibitive. Nevertheless, in all these instances of DBS ventures, the state has attempted to ensure that their television uses are not merely commercial and exploitative; even Britain has insisted on high quality programming!

As the chapters on Canada and the USA in this volume remind us, state interest in generating industrial growth in high tech areas is not a phenomenon unique to Europe. The USA's success with satellite is in large measure due to an early interest on the part of its government, the defence establishment and NASA in the development of space technology as part of the 'space race'. Canada has also experienced a large measure of departmental/military/governmental interest in the creation of an efficient and successful satellite industry and it is now benefiting from the fruits of that commitment in the same way as the American industry is.

Where the satellite services somehow fall outside specific national remits, for example, the Eutelsat, Intelsat and ASTRA ventures, governments have usually relied on their terrestrial broadcasting legislation to temper their excesses. In cases where advertising is not allowed on broadcasting services, there have been particular difficulties over the importation of advertising funded satellite delivered channels. One 'dilemma... is whether to apply the same standards and constraints on advertising to these media, and thus maintain a uniform code of practice, or to respond to market pressure to relax advertising restrictions'.[12]

This dilemma is partly founded on the realisation that governments cannot organise, fund and establish the whole array of new media technologies and services out of public funds. Few, if any, governments are able to fund the construction of technologically advanced cable systems, the construction and launch of satellite systems and the provision of programme and other services for these new technologies. Some governments, for example, the French, German and Japanese governments, have made some provisions for the new media but the enormous total financial cost of these projects — individually and together — has made some dependence on private funding almost inevitable.

Moreover, the current political preference in Europe, and elsewhere, for economic liberalisation and privatisation has ensured that many of these developments will take place within the private and not the public sphere so further reducing the state's control over broadcasting structures. The unwillingness and/or inability of governments to fund and establish the new media but, at the same time, to assent to their development further emphasises the degree to which these new media will become dependent on those forms of funding, notably advertising, sponsorship or subscription which places them firmly in the market place.

For some, the refusal on the part of the state to regulate and integrate the new media into both the public sphere and the national cultural and industrial context must be decried. But for others, it is a recognition of the critical part which advertising plays in the funding of both old and new media; it is also the clearest possible indication of the inability of states to continue to treat broadcasting as a scarce, controllable and national resource only. Satellite broadcasting removes, as it were, the straitjacket of European broadcasting where only a handful of organisations were permitted to control the air waves. It opens up the air waves and introduces 'public choice'.

For the entrepreneur, satellite broadcasting offers many opportunities; indeed, it would probably not exist in its current form without the entrepreneur. To understand why this is so, it is necessary to dwell briefly on the response of the television industry — incorporating everything from hardware to software manufacturers — to the prospect of the new media.

The availability of the new media has brought into existence a whole set of interests prepared to exploit the opportunities which they present. In the case of the USA, one can point to HBO; in the European context, to SKY, and so on. These are examples of the well tried tradition of the entrepreneur seeking new ways of satisfying a market. But with SKY, to name but one European service, this tradition is supplemented by some fundamental new factors which are specific to Europe. Amongst these is the belief, based on economic and financial realities, that the new media will not survive without the support of advertising and that one can satisfy the market without paying attention to cultural objectives. Both these factors are, as it were, outside the tradition of state regulated public service broadcasting.

Five key considerations inform the would-be satellite broadcaster. These include:

1. the knowledge that the new media will of necessity be commercial enterprises. Few governments are able to develop them with public funds. It follows that the new media will become dependent on advertising support or some form of subscription;

2. the knowledge that satellite services offer new outlets for advertisers. Across Europe, for example, advertising content on television is either strictly regulated or prohibited. Satellite broadcasting offers the prospect of new channels for advertisers and new ways of advertising products, including sponsorship;

3. the vision of Europe as a large market. Despite its fragmentation into political and economic units, Europe is a large market of over 350 million people. The fragmentation may create difficulties for the broadcaster interested in the European audience — will the British watch French or German programmes? — but it is still deemed to offer an opportunity to reach 120 m. TV households;

4. research findings[13] which suggest that there is a degree of dissatisfaction with the national broadcasters which means,

inevitably, that the satellite broadcaster should be able to attract that audience with the more 'exciting' package which will undoubtedly be on offer;

5. the knowledge that although advertising expenditure in some European countries is high and individuals spend a high proportion of their incomes on recreation and entertainment, it is low in many other countries and, on the whole, certainly much lower than in either the US or Japan. This suggests that there is room for benefiting from an increase in both dimensions.

Taken together, these points begin to highlight the motives of those who wish to enter the new competitive era. There is, it seems, a potential for exploiting the market and for profiting from that market. But there are also some major obstacles. Will all European states accept advertising funded, satellite delivered television channels? Will the national broadcasters face up to the challenge, say, by increasing their minutage of advertising and so reducing the attraction of satellite broadcasters? Can Europe be treated as a market given the immense differences in income, in the ability to gain access to the new media, in the degrees of satisfaction with existing broadcasting content and in wealth? Can a European lowest common denominator service satisfy wide cultural differences? How will governments view the prospect that with pan-European services, 'the potential exists to redistribute advertising revenue...from those viewing the programmes towards the countries originating transmissions'?[14] Will satellite broadcasting become nothing more than additional programme channels for the exploitation of home markets by those large media groups, Murdoch, Hachette, Bertelsmann, Springer which already have an interest in them? Finally, how will governments take to the view, expressed by Tydeman and Kelm that

> satellite communication from abroad, both direct to the household and to cable head ends, will provide viewing and advertising opportunities over which little or no effective controls can be exerted.[15]

This claim that governments will lose complete control is perhaps an oversimplification of a rather complex legal situation, particularly in the context of Europe where so many nations exist in such close proximity. Governments do retain some powers over cable operators who link individual households to satellite services. Indeed, in most countries across Europe, governments or regulatory bodies license cable operators and DBS programmers (see Ch.2). They could, therefore, refuse to license those operators which carry services not approved of, for example, because they carry advertising, carry obscene material or can threaten the national broadcasting services. Furthermore, governments may attempt to exercise powers granted to them under such acts as the 1960 Agreement on the Protection of Television Broadcasts and the 1965 Agreement for the Prevention of Broadcasting Transmitted from Stations outside National Boundaries (The Pirate Broadcasting Law). But there are moves to remove these national powers and to introduce more international broadcasting systems; such is the philosophy of both the Treaty of Rome and the EEC Directive 'Television Without Frontiers'. The latter has yet to obtain universal approval and widespread implementation.

In general, then, the prospect of satellite broadcasting does raise policy problems which call into question existing rules, for instance on advertising and foreign content, within specific states. As householders gain access to these new services via individual TVROs,the grip of the state becomes an anomaly. Yet there is a long way to go before the vision of unbounded cross-border television broadcasting becomes a reality. The satellites have to fly, the programme makers have to find the content to attract the European audience, the public has to be persuaded to part with its money, governments have to fail to act and national broadcasters have to concede defeat.

It is by no means clear that any of these important factors are certainties. Interestingly, it appears that the existing public service broadcasters are indeed gearing up to the future competitive era. Several such organisations intend to participate in satellite broadcasting. One finds, for example, the BBC, and various ITV companies in Britain playing a part in satellite services as programme providers. This suggests that the new media will only increase the importance of the existing broadcasting organisations as programme providers. Until the satellite services themselves become production organisations as well as transmitters of content, the dominance of the existing producers is likely to continue. Finally, one other trend seems to emerge from the present rapid developments and uses of the new media. Existing media organisations, most notably publishers, have also made inroads into the new media. The new media will probably fall into the hands of the present owners of national and international media concerns.

In the future, then, one is likely to observe further processes of alignment and re-alignment between the present strong players in the media entertainment field.

THE ORGANISATION OF THIS VOLUME

As the preceding sections have indicated, discussions of the new media are hedged by immense uncertainties. The two nations with extensive experience in these developments — Canada and the USA — offer contrasting visions of the future. The former paints a picture of a dependent country; the latter of a nation that is in an extremely powerful dominant media position.

For Europe as a whole, the future may be more akin to the Canadian experience than to the American one. No one European nation has yet gained dominance but the existing media players — particularly the English language services either via BBC/ITV or Sky Channel — are likely to prove strong contenders for the leadership. Will the Americans be far behind with their strength in programming?

One complicating feature as regards the European scene is that three culturally strong influences — France, Germany and Britain — have yet to cable their own countries. It could be, therefore, that the rate of new media development in those countries will be considerably slower than that experienced elsewhere and that this may create obstacles for media growth in Europe generally. Indeed, unless the pace of development quickens, the benefits of satellite broadcasting will be mainly available to central and northern Europe. These areas are already highly cabled and they can therefore make the most of the existing satellite delivered services. France, Germany and Britain have to decide on the proper balance between different new media developments. Should advanced cable systems be encouraged? Should DBS be centrally funded? What of medium power satellites? One lesson to be learned from the Canadian and American experience is that the satellite broadcasting–cable systems union remains a powerful and flexible one and far outweighs the benefits of DBS. Satellite broadcasting and cable offer an enormous capacity; DBS, in contrast, offers a limited number of channels and services. Hudson concludes her chapter in this book on the American experience as follows:

> While it appears that DBS is attractive for non-cabled rural areas, many of these viewers have already installed....antennas to pick up the channels being fed to cable systems. Even if more of these cable channels are scrambled, it is still likely to be more attractive to offer the same channels for direct reception by selling or leasing descramblers than to offer a separate satellite network for direct reception.

Whether other countries will take heed of these lessons remains to be seen.

The first two chapters of this volume contain material which is too often by-passed in considerations of the new media. The first chapter focuses on the technical aspects of satellite broadcasting, whilst the second covers many of the regulatory frameworks that confront all those concerned with this area. Both chapters relate their different contents to the issue at hand.

Other chapters in the book review developments in some major European nations — France, Germany and Britain — and across the Nordic countries where channels from outside are currently available and are proving to be popular amongst Nordic viewers.

Chapter 10 explores the rapid advances that are being made in Japan and the increasing use of DBS to provide high quality television pictures. Finally, Chapter 7 examines the background to current developments in the new media in Australia. Indeed, it goes so far as to draw parallels between it and the Canadian scene — a major parallel across continents.

No brief summary can provide an adequate description of the many influences that are currently changing the established means and structures of television communication. What does emerge very clearly though is that change *is* taking place, in varying degrees of rapidity, and that it is no longer possible to describe broadcasting systems as if they were static and immune from change. In the next decade, we shall be able to judge whether all this change has been for the better or for the worse.

REFERENCES

1. Negrine, R. 'Great Britain: The End of the Public Service Tradition' in Kuhn, R. (ed) *The Politics of Broadcasting*, Croom Helm, London, 1985.

2. Quoted in The Observer, London, 12.7.1987.

3. Negrine, R. (ed) *Cable Television and the Future of Broadcasting,* Croom Helm, 1985.

4. Hollins, T. *Beyond Broadcasting,* BRU, London, 1984.

5. Tunstall, J. *Communications Deregulation,* Blackwell, 1986, p. 129.

6. Ferguson, M. 'Broadcasting in a Colder Climate: Canada's Cautionary Tale' in *Political Quarterly,* Jan-Mar 1987, Vol. 58 No.1, pp. 41-2.

7. Ferguson, M. 'Broadcasting in a Colder Climate',p. 51, (emphasis supplied).

8. Tunstall, J. *Communications Deregulation,* p. 74-5.

9. See articles by Blevis, B. and Dua, M. in Singh, I. (ed) *Telecommunications in the Year 2000,* Ablex, New Jersey, 1983.

10. See Meisel, J. *Some Challenges of the Telecommunications Revolution,* Canada House Lecture Series, 22, 1983; and Sterling, C. (ed) *International Telecommunications and Information Policy,* NITA Report, Washington, 1984; *Cultures in Collision, The Interaction of Canadian and US Television Broadcast Policies,* Praeger, 1983.

11. Singleton, L. *Telecommunications in the Information Age,* Ballinger, USA, 1983, p. 85.

12. Tydeman, J. and Kelm, E.J. *New Media in Europe,* McGraw Hill, 1986, pp. 49-50.

13. Tydeman, J. and Kelm, E.J. *New Media in Europe,* 1986, p. 11.

14. Tydeman, J. and Kelm, E.J. *New Media in Europe,* 1986, p. 50.

15. Tydeman, J. and Kelm, E.J. *New Media in Europe,* 1986, p. 50.

1

Broadcasting by Satellite:Some Technical Considerations

Mark Williamson

THE SATELLITE ADVANTAGE

The first question to ask before embarking on a technical analysis of some new technology should always be ... why bother? If a new technology is to be worthy of discussion, it should present a discernible advantage over presently available technologies. Those that do have at least some chance of overcoming the political, financial and cultural pressures which haunt most of mankind's technical endeavours.

The overriding advantage of the communications satellite is its inherent ability to offer full regional, continental or global coverage at a stroke. Once a satellite is in position and operating, its signals may be received in any part of its coverage area, independent of the mountainous terrain, wastelands and ocean depths which prove such expensive barriers to terrestrial microwave or cable systems. This is why communications from space will always be a desirable option. Whether this option will prove viable for the specific application of broadcasting television programmes is another matter and must be analysed on an individual basis, for particular nations or groups of nations. This question of viability will be tackled in the forthcoming chapters, but before that we should investigate the workings of the technology which make this option possible.

DBS — A TECHNICAL DEFINITION

In its first two decades, communications by satellite has been limited in the way it provides a service to the recipient or 'end-user' by the technology available.

23

Satellites have been constrained by the amount of power they can radiate towards the receiving station, which has necessitated the provision of very large earth station antennas at a limited number of sites around the globe. From these 'gateway' stations the telecommunications signals, be they telephone, telex or television, are routed into the terrestrial networks and thence to the end-user.

Gradually, as the sensitivity of ground-receiving equipment has increased and the cost decreased, antenna sizes have been reduced and their geographical distribution has widened. It is now fairly common to see 3- or 5-metre-diameter dishes throughout our major cities and suburbs. The present position, however, is still not one of individual, direct reception from the satellite, there being usually a final terrestrial routing.

Over the years since DBS became technically feasible, the term Direct Broadcasting by Satellite has become increasingly vague. It now seems that almost any satellite system which delivers a TV signal to any antenna, up to the 3m-diameter antennas on hotel roofs say, is described as DBS. In light of this, it is useful to re-examine the reasoning behind the term 'direct broadcasting'.

'Direct' was intended to mean direct to the home and not to imply direct to the cable head-end, the point of reception and distribution in a cable system. TV 'broadcast' refers to the wide dissemination of a television service to the consumer and should be distinguished from TV distribution which is not intended to involve the consumer. TV distribution is the transfer of television programming from an originating source (say a TV company) to a point of distribution into the public network. The transmissions are not intended to be received directly by the public, as they are with direct broadcast, although they are often intercepted by radio and TV 'hams' or 'pirates', which has been the case in the USA for some time.

The distribution of programme materials between or within the broadcasting networks and operators has an analogue in the newspaper industry: newspapers are distributed to retail outlets (newsagents) who sell them to the public. There is no 'direct-broadcasting' of national daily newspapers to individual consumers (...yet). The interception of satellite signals classed as TV distribution is like stopping the van delivering to the newsagent and taking a newspaper! So, despite appearances, and common assumptions, TV distribution is not direct broadcasting.

What we may term 'true DBS' is what was intended by the delegates of the World Administrative Radio Conference of 1977

(WARC '77) when they developed the plan for DBS in Europe.[1] In a nutshell this postulated satellites with 5 channels of 27 MHz bandwidth, transmitting at power levels of 65dBW, requiring two or three hundred watts per channel of radiated power. This is now seen by many as an extremely wasteful and expensive way to broadcast TV, a view which has led to the appearance of what is called variously low-power DBS, medium-power DBS or quasi-DBS.The derivation of these terms is rooted in the technicalities of the satellite amplifiers or TWTAs (travelling wave tube amplifiers). At the time WARC was recommending powers of hundreds of watts for DBS, the most common output powers for satellite transponders were around the 10 or 20 watt level, requiring earth stations with 20 or 30 metre antennas to collect sufficient signal power. The WARC suggestion that individual reception should be possible using 90cm dishes drove the requirement for a new type of high-power satellite, using 200-300W TWTAs, thus distinguishing DBS satellites from other 'low-power' spacecraft used for general telecommunications.

The criticism now is that too pessimistic a view of technology advancement was taken in those early days, since it is already possible to receive TV transmissions from the European 'Eutelsat 1' communications satellite and 'Telecom 1', the French national satellite, with antennas as small as 1 metre in diameter from satellites with only 20W TWTAs.

However, neither satellite system was designed for DBS nor pretends to offer a DBS service. The transmissions received, therefore, may be termed 'quasi-DBS'. Although definitions can suffer from subjectivity, the output power of the satellite travelling wave tube is probably the best guide in defining the service. Most contemporary 'true-DBS' spacecraft have been designed to use tube powers greater than 200W (e.g. TDF-1, TV-SAT, Tele-X, and one of the four payloads on Olympus 1). Satellites using 100W tubes, or thereabouts, may be referred to as 'medium-power DBS' (e.g. the Japanese BS2's) and those operating on only about 50W, which seems to be the average anticipated for this last category, are 'low-power DBS' (e.g. Luxembourg's Astra).

The distinction between the various types of DBS (low power, quasi-, true DBS, etc.) is not truly fundamental. It is like trying to define 'hi-fi': when does it become medium- or low-fi? With DBS too, the important difference is 'quality' and, although there are many technical factors to take into account, quality depends to a large extent on the signal power (or 'power-density') received at the ground.

Lower-power satellites provide lower power-densities; smaller domestic antennas collect less signal power. Combining the two may provide acceptable quality most of the time but, for a receiver towards the edge of the coverage area in the middle of a heavy rainstorm, reception may be decidedly 'sub-standard'. In the same situation the high power satellite would generally have the power margin to maintain good quality reception.

DBS — THE SYSTEM

Apart from considerations of transmission power, the characteristic of fundamental importance to DBS is the signal path: from transmitting earth station to a satellite 36,000km distant and back to an individual receiving terminal. What makes direct broadcasting by satellite possible?

The signal, known as the carrier since it carries the TV picture, is amplified by the earth station equipment and passed through a filter, which confines the frequency of the radiated signal to its appointed bandwidth and reduces the possibility of interference with other similar signals. All communications signals are spread over a band of frequencies, the width of which depends upon the complexity of the signal: the bandwidth required for the DBS TV carrier is 27MHz.

Once amplified and filtered, the TV carrier is radiated from an earth station antenna towards the satellite. This is known as the 'uplink' path. It passes through the atmosphere and 36,000km of space before it reaches its destination, where it is again in need of amplification. It is collected by the spacecraft antenna and passed to equipment which filters the signal, changes its frequency for the 'downlink' leg, amplifies it and retransmits it. The amplification is performed by the previously mentioned travelling wave tube amplifiers (TWTAs).

The re-transmitted signal passes through the same distance of intervening space and atmosphere to the receiving ground terminal where, once again, it is amplified to give a usable input to the TV set which is the final link in the chain.

Fundamental requirements

There are two special and fundamental requirements for any satellite communications system which should be regarded as limited resources: a position in geostationary orbit and a band of radio frequencies to transmit and receive on.

Geostationary orbit is a special case in the family of orbits, in which a satellite appears to hang over roughly the same spot on the Earth's equator at all times. The stationary nature of the satellite relative to the Earth is what makes this orbit so important, since the earth station antenna does not have to move to track the satellite.

Incidentally the term 'geosynchronous' is often heard in place of 'geostationary' — in fact any orbit whose period of rotation is some multiple of the Earth's can be considered synchronous. Only the orbit whose period is exactly the same as the Earth's and whose plane lies in the plane of the equator is geostationary: its average height above the surface of the Earth is 35,786km (22,187 miles). This imaginary ring around the Earth was first brought to public attention by the science fiction writer Arthur C. Clarke in the October 1945 issue of *Wireless World*, and is thus sometimes referred to as a 'Clarke Orbit'.

As an example, Table 1.1 shows the plan of orbit positions for European direct broadcast satellites formulated by WARC in 1977. An interesting point about the orbital positions in Table 1 is that most of them have been chosen to be a number of degrees to the west of their respective country. This takes into account the effect of the 'eclipse season' when, around midnight (in the country concerned) for a number of days, the satellite orbits into the Earth's shadow where its solar arrays can produce no power. If the satellite was placed in an orbital position with the same meridian as the country, the direct TV service would have to cease as midnight approached. To counteract this, it is positioned to the west so that it is eclipsed in the early morning, when a 'close-down' is less important. Naturally, a 24-hour service could be provided, but in most cases it is considered uneconomic to carry the large mass of batteries sufficient to provide a high power DBS beam. The argument here is in favour of quasi-DBS satellites which, having lower power requirements, can maintain a service throughout an eclipse period. One major point which is patently obvious is that the satellites of many countries share the same nominal orbital position. Though at first this seems inadvisable, if not potentially dangerous, it should be noted that the geostationary orbit has a total

Table 1.1: European DBS orbital positions — WARC 1977

Orbital position   Country

Orbital position	Country
5° East	Cyprus, Denmark, Finland, Greece, Iceland, Nordic Countries, Norway ,Sweden and Turkey.
1° West	Bulgaria, Czechoslovakia, Poland, East Germany and Rumania.
7° West	Albania and Yugoslavia.
19° West	Austria, Belgium, France, Italy, West Germany, Luxembourg, Netherlands and Switzerland.
23° West	U.S.S.R.
31° West	Ireland, Portugal, Spain and United Kingdom.
37° West	Andorra, Liechtenstein, Monaco, San Marino, and Vatican.
44° West	U.S.S.R.

Notes: The 5 Nordic Countries (Denmark, Finland, Iceland, Norway, Sweden) will be allowed to use some of their channels for a joint regional services.

The U.S.S.R. has two orbital slots because over a large part of the globe it is the only country that both wants satellite channels and has the sovereign right to use them.

circumference of 265,000km (or 164,000 miles), which equates to some 736km (455 miles) per degree of longitude, so physical space is not too much of a problem.

Problems arise when the second fundamental requirement, *communications frequencies* are considered: if two satellites stationed at the same orbital position utilise the same frequencies, both satellites will receive each other's, as well as their own signals, leading to interference. Given that radio frequencies have to be shared out as carefully for space communications as with traditional radio and TV channels here on Earth, there is a limit to the number of different channels which can be beamed to a particular orbital position. This places a limit on the number of satellites sharing that position and makes both geostationary orbit space and radio frequency-space resources which must be conserved.

Part of the solution to this potential dilemma of interference is the allocation of different and widely-spaced transmit and receive radio frequencies to the various national DBS services. In addition, the signals are transmitted on one of two opposite hands of circular polarisation (right-hand or left-hand) which provides further isolation from nearby channel frequencies. Coupled with the angular separation of signal paths provided by differing orbital positions, frequency separation and polarisation allow a large number of non-interfering DBS services to operate, even within the constraints of limited orbit and frequency space.

Coverage

Practically every magazine article on DBS features a coverage diagram or 'footprint' produced by the satellite under consideration, comprising a simple ellipse or a pattern of concentric contours superimposed on an outline of the country (see Figures 1.1 and 1.2).[2] The footprint is the projection of a satellite's radio beam formed to a pre-determined size, shape and orientation by the spacecraft's antenna.

The contours may be likened to those on an ordnance survey map depicting a hill, since the centre contour borders the area within which the highest power-density is available and the others represent a progression to lower power-densities. Continuing the analogy, the centre of the beam is known as the 'peak', providing the 'peak power' within the footprint.

One common misconception is that signals cannot be received outside the coverage area delineated by the simple ellipses in Figure 1.1. In fact these ellipses usually represent the contour around which the power from the satellite is half what it is at the peak of the beam. The area within this contour (in technical circles known as the -3dB contour) is termed the *beam area*, but of course the definition only applies if it really is a half-power contour — anyone can draw an ellipse on a map!

The point of this is that there is still an appreciable level of power outside the beam area and signals can still be received, albeit with a larger antenna dish than nearer the centre. There is, therefore, no physical cut-off point to a DBS service, or any other

Figure 1.1: European satellite footprints

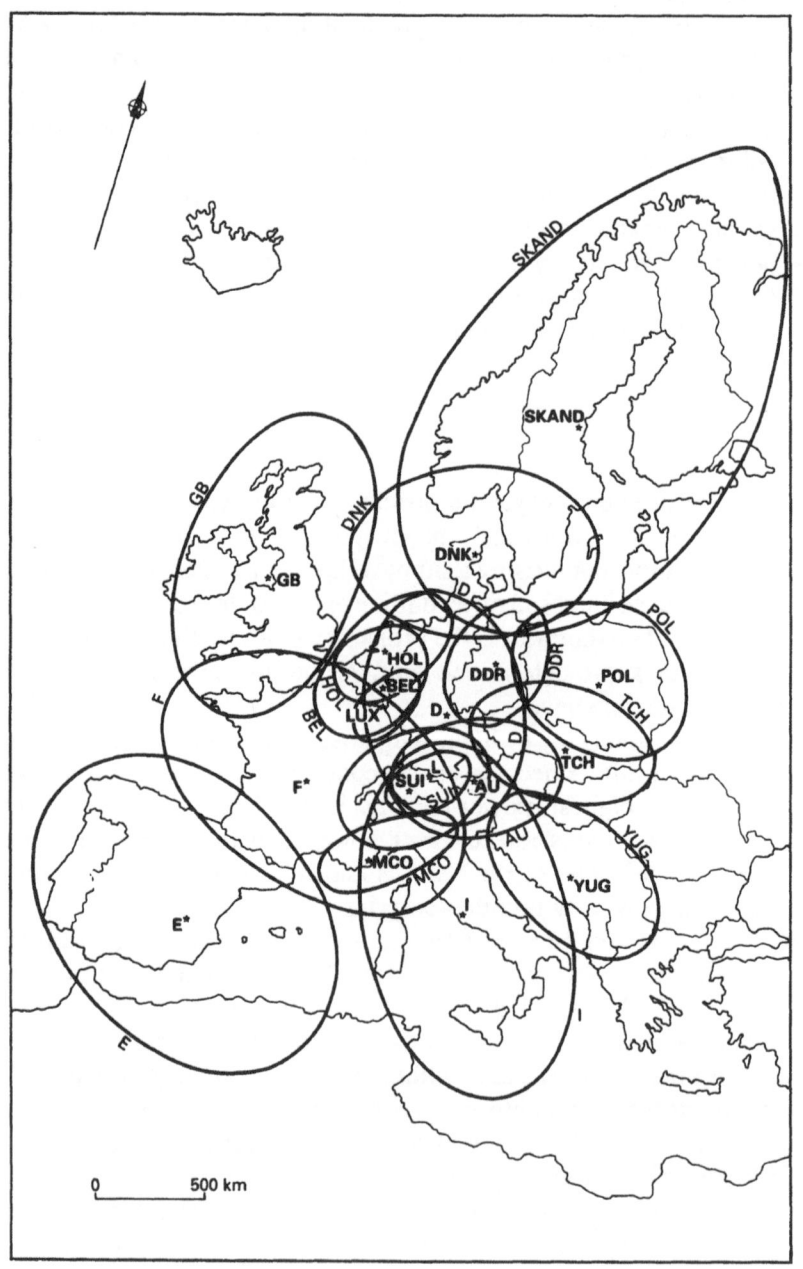

Figure 1.2: UK WARC footprint

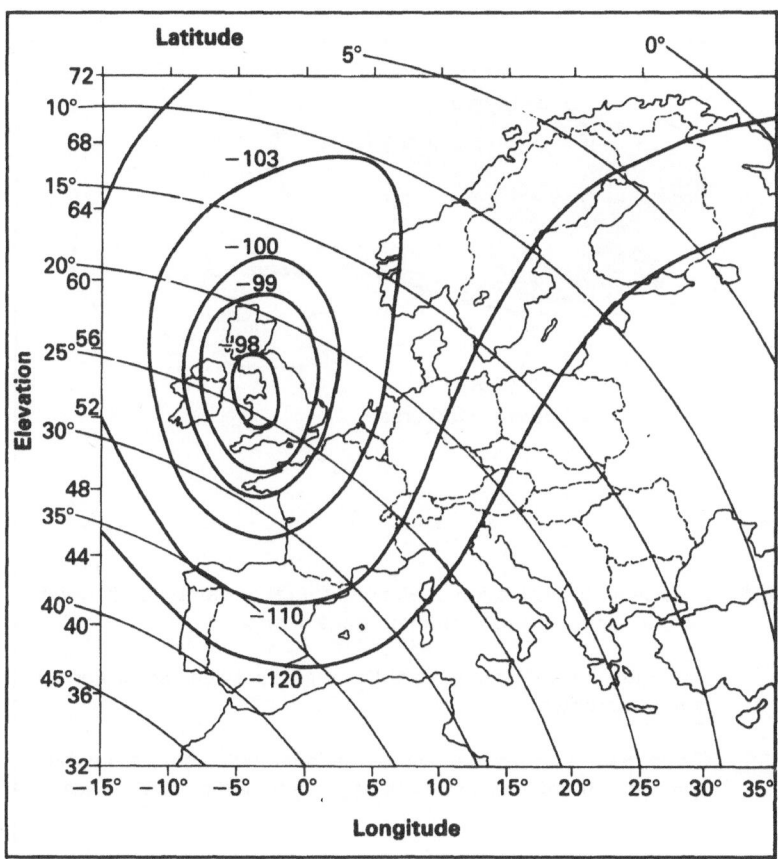

satellite telecommunications service, and coverage can be extended by recourse to more sensitive, and more expensive, equipment.

In an attempt to obviate confusion between terms which were being used interchangeably, WARC '77 made a further distinction between 'service area' and 'coverage area'. Figure 1.3 shows their relationship.[3]

The *service area* is defined as the area within which the administration originating a service can demand protection against interference from other transmissions. A receiver in the service area should therefore be able to receive an interference-free signal from that country's satellite. It is assumed that the service area lies inside the political boundary of the country, but WARC is not specific on

31

Figure 1.3: The relationship between the service area and coverage areas of two adjacent countries.

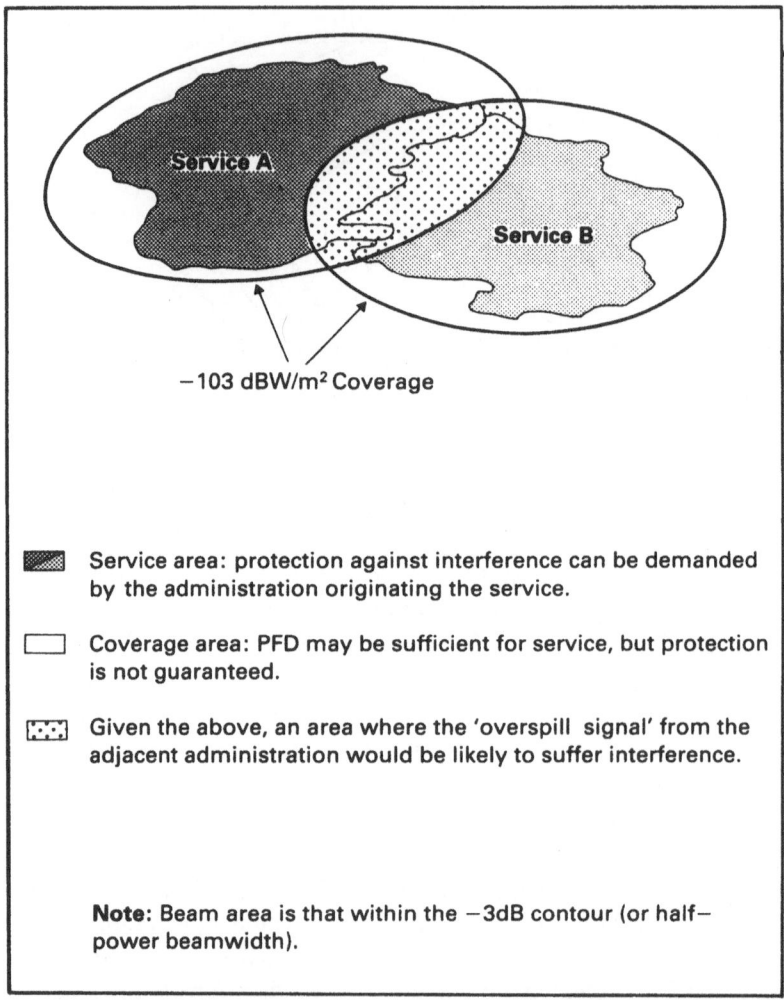

Service area: protection against interference can be demanded by the administration originating the service.

Coverage area: PFD may be sufficient for service, but protection is not guaranteed.

Given the above, an area where the 'overspill signal' from the adjacent administration would be likely to suffer interference.

Note: Beam area is that within the −3dB contour (or half-power beamwidth).

the point, recognising that some political boundaries are subject to dispute.

The *coverage area* is defined as the area over which the signal power level may be sufficient for a DBS service but in which there is no guaranteed protection. The signal power would in this case be sufficiently high to permit the desired quality of reception, but only in the absence of interference.

Although such definitions of area appear at first sight academic, it is to technicalities such as these that the legal profession may refer in future cases of dispute, where one country's services are found to be interfering with another's.

The domestic antenna

Probably the most obvious component of a DBS system to the casual observer is the 'dish' or parabolic communications antenna. This is the item which collects the signal radiated from the satellite and channels it to the receiving equipment adjacent to or, in future, within the TV set.

The size of the antenna specified in the WARC Plan has greater ramifications for the system than are at first evident. The CCIR (International Radio Consultative Committee) originally proposed that a 75cm dish should be the standard, obviously having well in mind the cost and environmental impact of the antenna. The diameter was, however, increased to 90cm by the WARC committee in response to recommendations from the US delegation, who pointed out the importance of improving the total capacity of the frequency band and the geostationary orbit.

Their argument rested on the relationship between the size of an antenna and its *beamwidth*, which is defined for a satellite antenna as the width of the beam of radiation shaped by the parabolic surface and extending to the points where the radiated power is half its peak value. It is often called the half-power beamwidth and its projection on the Earth's surface is the above-mentioned half-power contour. Although the domestic antenna does not transmit (hence the acronym TVRO, for Television Receive Only), it has a *receive beamwidth* which can be thought of as a sky-pointing cone into which radiation from a satellite may fall.

A large antenna has a narrow beamwidth and must be pointed accurately at the satellite, but a narrow 'cone' means there is less chance of receiving unwanted signals from adjacent satellites. Conversely, a smaller antenna with a wider beamwidth is easier to align with the satellite but increases the risk of interference.

The increase in the size of the home DBS antenna from 75cm to 90cm narrowed the beamwidth and allowed the nominal orbital spacing of broadcasting satellites to be reduced from the 7.5°, proposed in earlier planning exercises, to the 6° now adopted. The resultant 25% increase in the number of available orbital positions was one of the factors which made the WARC Plan possible in the first place.

It has been suggested that 90cm amounts to 'overkill', since all orbital positions for DBS will not be occupied in the foreseeable future, so 30cm 'window-sill' antennas have been proposed. However, an additional characteristic of small antennas is that their gain is lower: they receive less of the satellite's radiated power because of their small area. In fact the 30cm dish has a gain of only about a tenth of the 90cm antenna, which would mean that either the satellite would have to transmit ten times the power, or the receiver electronics would have to be ten times as sensitive to enable the same quality of TV picture to be received. Common sense shows that it would be far cheaper to increase the size of a mass-produced antenna rather than enhance the design of the domestic receiver or the satellite. Moreover, if low power DBS becomes the norm, the larger, higher gain antennas will be required anyway. One may well see 30cm DBS antennas on sale in future, but it is likely that they will provide inferior reception at a 'budget' price.

A development which may please the environmentalists who fear the prospect of antenna dishes on every rooftop, is that of the 'flat-plate' antenna.[4] Much experimental work has been conducted on microwave antennas comprising a number of radiating elements mounted on a flat surface, which can be electrically coupled to form a beam. For DBS the plate could be mounted on a house wall or installed as part of a pitched roof. The beam could be 'steered' electronically to point towards a number of satellites in different orbital positions. Much of the work so far has been associated with military projects and, until recently, flat-plate antennas could not be made economically enough for the consumer market, but, thanks to recent research, they may eventually replace the 'dish-antennas' seen by some as a potential eyesore.

Coding the Signal

Although a colour television picture is something most people take for granted these days, the manipulation of the radio signal which carries the picture is an exceedingly complex process. A 'state-of-the-art' comparison between TV technology and space technology exemplifies this: although it was possible in 1969 to land a manned spacecraft on the Moon, the 'live' TV pictures returned were indistinct, ghostly and transmitted in black-and-white.

The three world TV transmission standards used today (in a variety of sub-forms) PAL, SECAM and NTSC, were developed in the 1950s to be compatible with existing monochrome receivers. Because of this the designs were a compromise, but the advent of DBS offered the opportunity to develop a completely new system which would give higher definition pictures and have the potential for further improvements in the years to come. It was with this in mind that the UK Independent Broadcasting Authority (IBA) developed the system currently being implemented for European DBS: Multiplexed Analogue Component (MAC).

A colour TV picture comprises two different sets of picture-information, details of brightness (luminance) and of colour (chrominance), whereas the monochrome systems had only luminance. With the terrestrial colour systems mentioned above the chrominance and luminance information is interleaved within the frequency bandwidth of the signal in an ingenious, if complex, way. One of the drawbacks to this mixing of colour and brightness is that the two information-sets can interfere. One effect is the 'colour flashing' seen on clothing (historically newsreaders' jackets) with a closely-lined pattern or check. The MAC system virtually eliminates such problems by electronically compressing the luminance and chrominance information and transmitting them sequentially rather than simultaneously. The signals are expanded and used to create a picture in the receiver.

The main drawback is that MAC is not compatible with current TV receivers and will require the addition of an adapter or converter. It was for this reason that the contending system proposed by the BBC, Extended PAL, was made compatible with standard PAL receivers.[5] This was, of course, the same philosophy used for the move from monochrome to colour and was criticised for being 'old-fashioned' and restrictive.

In November 1982 the argument was settled in the UK by the report of the Part Committee which recommended MAC to the UK government, since it was considered to be the best technical solution and have the better chance for European unity on standards. At the time, the BBC had a valid counter-argument that the initiation of MAC would leave the UK isolated in Europe, since France and Germany, already building their DBS satellites, had entered into a bilateral agreement to use versions of SECAM and

PAL. This would severely restrict the market for manufacturers of DBS receiving equipment.

However, Germany and France were later persuaded to adopt the MAC system, largely because patents governing the manufacture of receivers exclude Japanese imports and give European manufacturers time to establish themselves. There would be no such barrier to the import of Japanese PAL and SECAM systems. But technical standards are rarely this clear cut: the UK had originally adopted a version of MAC called C-MAC, but, after great deliberation, changed to D-MAC in 1987. This version is still, however, not compatible with that of Germany and France who have chosen D2-MAC, a reduced bandwidth version which is compatible with the lower bandwidth cable TV links. The drawback here is that, whereas C-MAC has sufficient bandwidth for eight digital data or sound channels, D2-MAC has only four. This means that D2-MAC is far less suited to future developments in television such as the large screen, high definition pictures planned for the 1990's which will have perhaps 1250 lines as opposed to the present 625 currently standard throughout Europe. The D-MAC bandwidth allows fewer sound channels than C-MAC, but retains the disadvantage of being too great for cable networks.

Agreement on technical standards is invariably fraught with problems: take the comparable field of domestic video, which offered VHS, Betamax and V2000, or the more historic non-compatibility of railway gauges. An exception to the rule was the standard Philips audio cassette. DBS transmission standards seem destined to remain like currency: conversion is possible but a loss is inevitable.

DBS — Regulatory and Political Aspects

WARC '77 was convened by the International Telecommunications Union (ITU) and held at the International Conference Centre in Geneva from 10 January to 13 February 1977. Its main goal was the establishment of a plan for the broadcasting satellite service (BSS) in the 12GHz (downlink) band.[6]

The cast of WARC '77 was truly international: it was attended by more than 600 delegates from 111 countries, as well as officials from the ITU and observers from a number of international organisations, including the European Broadcasting Union (EBU). The origin of the Conference dates back to 1971, when the WARC

Figure 1.4: WARC regions 1, 2 and 3.

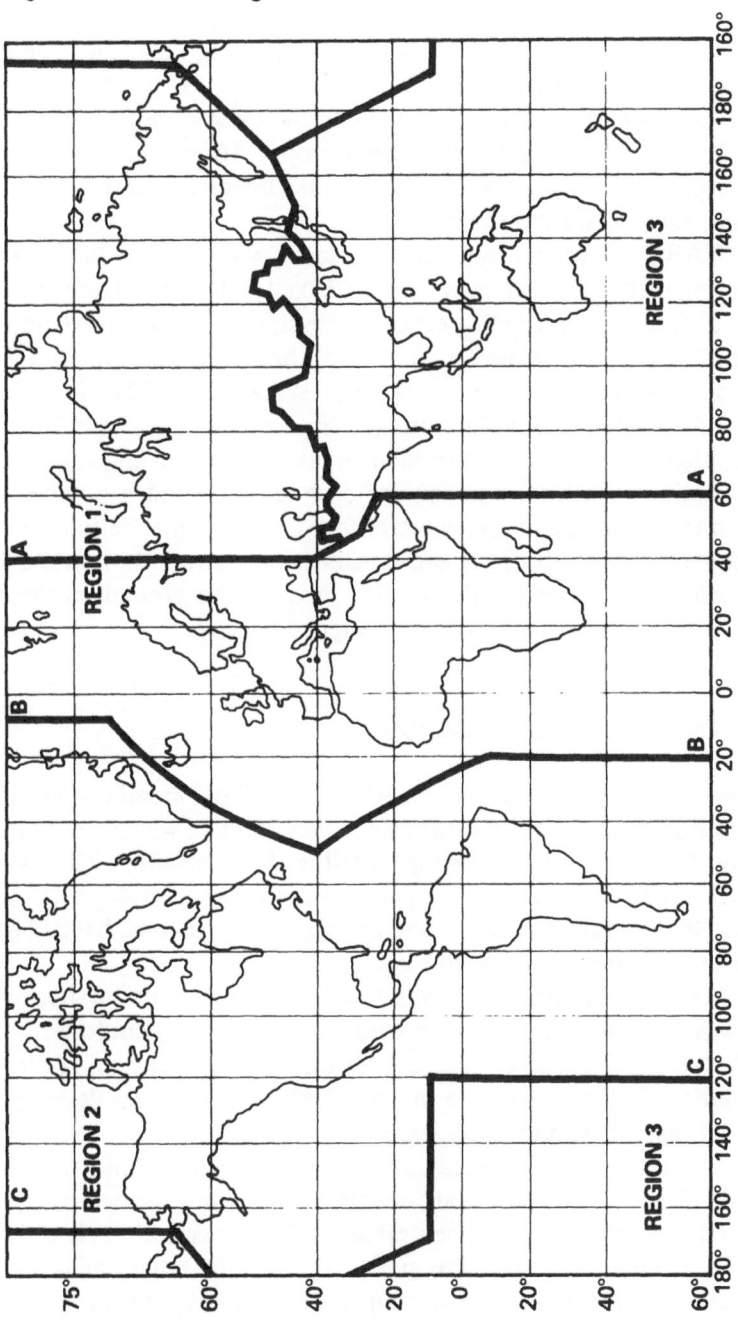

for Space Telecommunications (WARC-ST) made the first allocation of frequency bands to the broadcasting satellite service. The allocations differed slightly depending on the region concerned, but were between 11 and 13GHz. For administrative convenience, the Earth is divided into the three regions illustrated by Figure 1.4: Region 1 includes Europe, Africa, the USSR and Mongolia; Region 2 the Americas and Greenland; and Region 3 Asia, Australasia and the Pacific.

WARC-ST recognised that problems would arise due to the sharing of frequency bands with the Fixed-Satellite Service (trunk telephony, etc) and indicated the necessity of a world or regional plan for the BSS. It made no attempt, however, to formulate the plan, leaving it for a future WARC: WARC-BS-1977. The plan would involve the definition and integration of a number of technical characteristics for each country. These would include a geostationary orbital position for each satellite and a number of channels to transmit on, which would require the definition of frequencies, polarisations and transmission powers. It would also be necessary to define the service area on the ground so that a satellite antenna beam could be designed to cover it.

WARC '77: attitudes to the plan

The broadcasting organisations which were EBU members were in favour of a plan from the start, since they felt this was the only fair way to initiate and develop satellite broadcasting. This was particularly important for the poorer nations, who would be ousted from the field if frequencies and orbital slots were to be handed out on a first-come, first-served basis.

In fact, the 12GHz band was chosen for DBS downlink because it was the band for which contemporary technology could provide relatively inexpensive receivers for individual reception. If direct satellite TV was ever to be available to the less developed countries, this had to be the case.

The European terrestrial telecommunications administrations were also in favour of a plan, mainly because they needed to know which frequencies would be usable for the terrestrial services they were planning to install. For similar reasons the other countries in Regions 1 and 3 were in agreement, particularly the developing nations who were anxious that their future requirements should be recognised.

Opinions in Region 2, however, were not so concordant. Some supported the idea of an immediate and detailed plan, but most, led by the United States, preferred an evolutionary plan where assignments would be made on an *ad hoc* basis. Services would be assigned frequencies as and when required, so that full advantage could be taken of state-of-the-art technology. This would be tempered, however, by a guaranteed access to the two limited resources of geostationary orbit space and the frequency spectrum, and would form the basis for non-interference solutions. The sharing of the band with the Fixed-Satellite Service would make this doubly necessary.

The result of this difference of opinion was that WARC '77 included only Regions 1 and 3 in its detailed deliberations (Region 2's plan was formulated some six years later in 1983). Even this, as it turned out, was no easy task. After all, who could expect 600 delegates from 111 different countries to agree on anything! Naturally enough, the large number of competing national requirements for severely limited natural resources led to conflict. Unfortunately few people had the technical background required to solve these problems by devising a plan which everyone would adhere to. Individual national requirements for channel frequencies and orbital positions had been collected and published by the ITU prior to the Conference. As negotiations progressed the requirements were amended and, practically every day, the International Frequency Registration Board (IFRB) issued a new set of updated requirements. Sometimes details would be altered because the delegates, who were learning more about the problems, simply changed their minds.

Even before the Conference began, it was obvious that some requirements were excessive and would lead to higher levels of interference than could be allowed. Some delegations wanted more channels than could be accommodated within the limited bandwidth assigned to the Service and some requirements for power flux density (a measure of the signal power on the ground) would have led to serious interference problems. With so many nations in Europe clustered closely together, their first preference was for the same orbital position which, in much of Europe's case, was 19°W. This meant that those delegations which argued less persistently were relegated to less favourable positions like the UK's allocation of 31°W. The drawbacks of this position include a reduced elevation of the satellite above the horizon and an increased obliqueness of the beam footprint on the Earth's surface.

Basic principles

The underlying principles of the DBS plan include two fundamental provisions: a broadcast TV signal direct to the home (individual reception) and national coverage for most participating countries.

In pursuance of these principles each country was allocated an orbital position and a number of channels with a particular hand of polarisation. Most countries in Region 1 (Europe and Africa) received five channels, while the countries of Region 3 (Asia and Australasia) received either four or five.

With their characteristic flair for equality, the WARC committees made equivalent assignments to all countries, however small: in Europe, for instance, Andorra, Monaco, Liechtenstein, San Marino and the Vatican State received allocations equal to those of France, Germany, Switzerland and the UK. Further afield, assignments were made to Singapore and the small islands in the Indian and Pacific Oceans. It is, at present, difficult to visualise five DBS channels emanating from San Marino, for example, but WARC '77 was a plan for the future — and which of us can foretell the future?

Perhaps surprisingly, the Vatican State already runs one of the most extensive communications systems in the world. Radio Vaticana, designed and installed in 1931 by Guglielmo Marconi himself, now broadcasts 24 hours a day in 34 languages worldwide, so it may well find a use for its DBS allocations long before some of the more major European nations.

Most countries have been allocated one beam for one service area, but certain countries are too large and have been divided into several service areas. A few examples are listed below:

- 2 areas; Algeria, Libya, Saudi Arabia, Afghanistan, Malaysia, etc;
- 3 areas; Sudan;
- 5 areas; Pakistan (with 3 channels in the largest area and 2 in the others) and Indonesia;
- 6 areas; Australia (with 6 channels per area);

In addition, India is divided into 12 areas, the USSR into 21 and China into 35, which all goes to make the planning for Europe look trivial!

One further anomaly of the Plan is represented by Iceland which has been allocated two orbital positions. One 'slot' is well to the west and carries five channels, while the other is the position shared by all the Nordic countries, and provides five other channels for the coverage of Iceland and the Faroe Islands. Within the Plan, a number of examples of 'legal spillover' were recognised. Amongst them were a number of so-called 'superbeams' which cover a group of neighbouring countries or greatly exceed the national frontiers of one particular country. For example, a Nordic superbeam covers Denmark, Finland, Norway and Sweden with 8 channels (2 for each country). By way of compensation, the number of national channels for each of the Nordic countries has been reduced to three.

The request for a similar superbeam for the German-speaking countries (West Germany, Austria and Switzerland) was denied, with the allocation being limited to the standard five national channels for each. The coverage areas of the beams were, however, slightly extended as a consolation.

In addition to four narrow-coverage beams, the Vatican State has been allocated a channel covering practically the whole of Italy—perhaps some idea of the Vatican's future intentions can be gauged from this. In fact the Vatican's orbital position of 37°W affords an excellent view of the South American continent and the Eastern USA with their resident Catholic populations. The future expansion of Radio Vaticana's influence will be an interesting 'window to watch'.

Lastly, but on a different religious polarisation, the Moslem countries have agreed to the broadcasting of a joint Islamic religious programme. To provide for this service, the coverage of a channel from each of Saudi Arabia's and Syria's allocations has been extended. Other countries will use one of their national channels for this programme.

Disagreement

Although the Conference was very largely a technical success, it would have been optimistic in the extreme to expect no political disagreement in a forum of 111 participating countries. One fundamental question was discussed at great length without the appearance of a clear solution: the sovereignty of the geostationary orbit.

The countries which lie about the Earth's equator developed a number of legal arguments to assert their rights to exercise control over the portion of geostationary orbit directly above their territory. They wished to apply national legislation to their 'geostationary orbit space' in the same manner as the already recognised 'air-space'. Under these conditions a broadcasting satellite in full conformity with the plan could not be brought into service above an equatorial country without that country's express agreement.

Naturally the opposition sought to show that geostationary orbit was a part of outer space and thus escaped all claims of national sovereignty. Although the matter was discussed at the 1985 WARC-ORB conference on the use of geostationary orbit, the implementation of any formal plan will have to await the second half of the conference, due in 1988, which will decide upon the actual planning process. One can only hope that near-Earth space is not divided into national territories at some future meeting in a similar manner to the division of the continent of Antarctica.

RARC 1983

It was against the background of WARCs '77, and '79 where the Plan was ratified by a plenary assembly, that RARC '83, the Regional Administrative Radio Conference, opened its doors to the 47 countries of Region 2. The meeting was not without discord.

Prior to WARC '79, the US had received a good deal of criticism, which they considered unfounded, that they were not prepared politically for the Conference. Their lack of willingness to accept an organised Plan for Region 2 in 1977 may have engendered this view, but at RARC '83 the boot was on the other foot. It was considered that the Final Acts adopted at WARC '77 contained 'A rigid spectrum/orbit plan based on conservative technological assumptions'. The Federal Communications Commission (FCC) maintained that it was very difficult to make modifications to the Plan, because of its rigidity, but this so-called rigidity was seen on Europe's side of the Atlantic as the Plan's cohesive strength. The disagreements of WARC '77 were resurrected.

Accordingly, the FCC encouraged RARC '83 to go beyond the 1977 parameters due to advances in computer modelling techniques and satellite technology generally. The US delegates did, however, concede that really the United States would prefer to have no Plan at all and introduce communications systems as they occurred. But since the rest of the world wanted a plan, a Plan there would be.

The United States considered that the allocation of frequencies and orbital positions would be a relatively straightforward task for Region 2 compared to that for Region 1, whose countries had been obliged to reduce their initial requirements to give their production of a plan at least a fighting chance. Region 2, after all, has a relatively small number of countries and covers a comparatively wide range of longitudes (and therefore orbital slots). The opinion was, therefore, that it should be unnecessary to assess every possible combination of frequency and orbital position. The rather insular assumption inherent in this view was that the other countries in Region 2 would not request more than a handful of channels between them. Naturally this was not the case: Paraguay and Uruguay requested 18 channels each!

The question asked, however, was: when will they use their resource? The Federal Advisory Committee studied the matter and concluded that these countries will never implement a Broadcast Satellite System, because it will always be cheaper for them to use terrestrial networks. The world will have to wait and see.

When the Americans arrived at RARC '83, they found that many administrations had greatly increased the channel-requirements previously submitted to the IFRB — the total number had doubled from around 1,000 to approximately 2,000.[8] This level of competition must have surprised the American delegation, but luckily the world of the ITU radio conference is a democratic one and every over-subscribed country had to reduce its requirements accordingly.

Despite the United State's published wish to confine RARC '83 to the technicalities, they must have realised that no such conference could remain non-political: sharing out the natural resources of geostationary orbital space and frequency spectrum evoke the same feelings of nationalism as do shares in fishing grounds in coastal waters or the possibilities of lucrative drilling tracts in Antarctica.

DBS — THE SATELLITES

A satellite is, by its very nature, unique since no other product of engineering suffers such enforced isolation, 36,000km from the nearest spare part or maintenance technician. Reliability is therefore of prime importance, and as such is a branch of engineering in its own right, predominantly concerned with failure statistics. This high reliability requirement leads to the most exacting standards in choice of materials and components, electronic and mechanical design, manufacture and testing. The overall design philosophy is, perhaps surprisingly, conservative in that the number of entirely new developments is kept to a minimum, so that the basic spacecraft is invariably based on an existing design incorporating flight-proven sub-systems. There are, however, exceptions to the rule.

Earlier we pursued a discussion of the definition of DBS based on the power levels provided by the satellite showing that low-power and quasi-DBS spacecraft are not significantly different from ordinary telecommunications spacecraft since they utilise relatively low powered amplifiers. Direct broadcast satellites, however, represent a deviation from the norm in that they utilise relatively high transmission powers. In fact, all the current European DBS satellites, the French TDF-1, the German TV-SAT and the Swedish Tele-X, are in the high-power category.

The current move, from the preponderance of 10 and 20W travelling wave tubes (TWTs) on existing spacecraft, to DBS tubes of up to 260W represents a quantum leap in tube-power and a departure from the traditional conservatism.[9] The Olympus 1 demonstration satellite will carry a number of 230W tubes as part of a European experimental DBS payload, but it seems likely to be beaten into orbit by the operational TDF-1 and TV-SAT, meaning that 230W tubes will be in-orbit-tested in anger.

Although the TWT is one of the least reliable spacecraft components, due to its tremendous complexity, the reliability is only relative to other components and tubes should not be branded as 'unreliable'. As a result of 30 million hours of performance testing on a particular type of low-power TWT, the Hughes company calculate the probability of failure for this type of tube as once every 5 million hours (or 570 years!). This, of course, does not imply that the unit will work for 570 years — it simply states that the probability of failure within its lifetime is very low (e.g. every 57th tube may fail within its 10 year lifetime). Beyond this, the statistically significant data base becomes rather sketchy since the tube manufacturers guard their results like state secrets, and it is not clear whether high-power TWTs will be less reliable than low-power tubes. In fact both the major European tube manufacturers have experienced problems with the development of associated equipment for TV-SAT and TDF, contributing to delays in the satellite launches.

However, where possible, redundant components ('spares') are carried on board a satellite, as part of a policy to circumvent random failures. Some are switched on automatically in the event of a failure and others are commanded by ground control. A common redundancy scheme for TWTs is to fit 'three for two', whereby each pair of tubes has a spare which can replace either of the prime tubes.

In terms of design of their communications payloads, all current DBS satellites are configured for services to a single originating country and could not be used for another country's services. They are all unique: their antennas produce footprints of different shapes and sizes; they transmit and receive on different radio frequencies and one of two opposite polarisations.

One attribute the DBS satellites scheduled for launch in the next year or two have in common, however, is that they exist only because they are not truly commercial ventures. They are all funded by either international or governmental organisations as long-term, somewhat experimental, investments. Commercial DBS is a high risk business due to the massive initial investment in hardware and systems for virtually unknown financial returns.

A possible alternative to this high-risk solution is for several different providers to share one or more satellites for an initial service on a reduced number of channels, or for several countries to share a back-up satellite configured for use by any of them. Transponders able to switch between the various channel frequencies would be carried, along with steerable antennas able to point at whichever country desired a service. Payloads featuring these attributes are termed 'reconfigurable'. Extra station-keeping fuel would be carried to enable a change of orbital position as and when necessary. The technology is available (and some of it will be demonstrated by the Olympus 1 satellite), but a requirement for the technology and the associated political agreement have yet to be secured.

As the insurance community becomes better informed on the technicalities of communications satellites, matters such as redundancy and reliability take a higher profile. The main difficulty in any prognosis concerning the space industry is the lack of a meaningful statistical sample. The industry is still too immature, compared with the marine and aviation sectors, to be able to make really reliable predictions of failure or success. It will not be long, however, at the rate of development foreseen for the next decade, before the subject of risk analysis begins to take on a more scientific image as opposed to the 'art' of informed extrapolation that predominates today.

MARK WILLIAMSON

THE LAUNCHER MARKET

Following a number of satellite failures prior to the January 1986 Challenger accident, insurance rates exceeded 30% for launch and initial in-orbit coverage, assuming insurance was available at all. This was a direct result of a loss ratio of 200% suffered by the space insurance community over the period 1977 to 1985: $900 million was paid out in claims against receipts of about $445 million. Taking the period 1983-1985 alone, the loss ratio was 330%. Naturally enough the failures of four separate western launch systems in 1986 — the Space Shuttle, the Titan and Delta launchers, and the European Ariane did little to increase the optimism of insurance brokers and underwriters.

Even when Shuttle flights resume, the emphasis on commercial payloads, such as DBS satellites, will be gone, priority being given to military and space station payloads. Thus plans are afoot to develop alternative expendable launch vehicles (ELVs) to replace the dwindling stock of American launchers, but it is unlikely that these will be available much before the end of the decade. Europe's Ariane is fully booked into 1990 and, of course, any further failures will result in additional delays to those launches.

However, the availability of launchers does not end with Europe and the USA. The Chinese are offering their Long March 3 booster for commercial launches and have attracted, amongst others, Western Union Telegraph Co. who intend to launch their Westar 6-S spacecraft on the Long March. Japan are not far behind with their H-Series boosters. The first stage of the H-I, first tested in August 1986, is effectively an American Delta, but the second and third stages are of Japanese design. The H-I allows the launch of only a 550 kg spacecraft into geostationary orbit, but by 1992 the much larger H-II should be available to orbit 2000 kg payloads, thus competing with Ariane and its equivalents.

Several analogies, from wagon-trains opening up the west to commercial trans-Atlantic airlines, have been applied with some validity to the current period of space utilisation. Failures in the early phases of any technological endeavour are to be expected and there will be launch vehicle failures in future, but as confidence increases with an increase in reliability, the business and insurance communities will begin to think of rocket launches much as they do Concorde flights today.

CONCLUSION

The 1977 World Administrative Radio Conference on satellite broadcasting produced a major document for world communications and its success as the 'incubator' for the DBS Plan is now part of the history of space telecommunications. The birth of DBS has not, however, been devoid of pain: companies have been formed and disbanded; satellites have been proposed and rejected; money has been spent and wasted. The contents of this book will help the reader to decide whether DBS is to become a technological tool of tomorrow or just another brave failure. For now that the technology is available, in all its many guises, technology is no longer the question. High power travelling wave tubes and parabolic antennas are here to stay, but once the first generation of DBS satellites have been retired will others take their place? The questions now are phrased in cultural, political and financial terms.

This chapter opened with the statement that if a new technology is to be worthy of discussion, it should present obvious advantages. Whilst DBS does so in certain circumstances, cost and other considerations have to be met. For instance, many countries already have extensive cable networks and perhaps for them the promise of DBS is not of sufficient interest to initiate a system. However, it is extremely unlikely that the mountain communities of Switzerland or Wales, or the island communities of Indonesia or Greece, or the desert communities of Australia or Saudi Arabia will ever be completely cabled, so in these cases DBS is the ideal method for broadcasting television. Where the service area of a country has a mixture of both urban areas and sparsely populated terrain the probability is that DBS and cable will operate together. Indeed most modern tower blocks and some towns and cities are already cabled, often due to a local restriction on terrestrial TV antennas. For these reasons DBS and cable should not be seen as direct competitors: the two can be complementary.

It is argued that too little effort has been spent on the product side of DBS. How much more television do we want and what sort of programming would we like? The DBS Plan offers the UK the potential of five extra channels, which would more than double the choice of broadcast output available. As the advent of DBS draws nearer, questions are being asked about the quality of the programming, especially if the major source of income for the channel is advertising revenue, the argument being that there is too little advertising money to fund the increased number of channels.

Radio transmissions do not observe the restrictions of international borders and satellites offer the ultimate potential for the broadcast of political, or simply cultural, propaganda. Russian satellite TV is already easily received in parts of Europe; European programming will soon be spilling into the countries of the Eastern Bloc. Within Europe itself, this 'overspill' could lead to the reception of what one nation considers pornography from another which considers it merely a vehicle for advertising.

If, in order to improve profit margins in an industry of uncertain returns, corners are cut in the design and operation of DBS satellite and ground-based equipment, interference between services may well result. This introduces a legal aspect to the fray: if just one country initiates a service which bends the rules of WARC '77 and creates a potential for interference, the WARC Plan is at risk. Another service arriving five or six years later could find its 'legal' service in jeopardy even within its national boundary. If DBS systems are to proliferate, the policemen of the ITU will have to lay down the law.

The remainder of this volume will indicate the viability of DBS for individual countries and the likelihood of its development in the coming decade.

REFERENCES

1. Final Acts of the World Broadcasting Satellite Administrative Radio Conference, International Telecommunications Union, Geneva, 1977.
2. Williamson M., 'Making Footprints on Europe', *Satellite and Cable Television News*, No. 16, Feb 1984, pp. 52-6.
3. Williamson M., 'Behind the Scenes at WARC '77', *Satellite and Cable Television News*, No. 24, Oct 1984, pp. 45-54.
4. Jones R.E., 'Flattening the Dish Theory', *Satellite and Cable Television News*, No. 22, Aug 1984, pp. 54-6.
5. Fox B., 'MAC versus Extended PAL', *Professional Video*, Sept 1982, pp. 46-50.
6. Brown A. & Mertens H., 'The Work of the Conference on Satellite Broadcasting', *EBU Review*, April 1977, pp. 60-7.
7. Kachmar M., 'US Seeking to Launch DBS at RARC '83', *Microwaves & RF*, June 1983, pp. 28-9.
8. Kachmar M, 'RARC '83 Resolution', *Microwaves & RF*, Aug 1983, p. 29.
9. Williamson M., 'The Physics of Space Tubes', *Physics in Technology*, Vol. 17, No. 5, Sept 1986, pp. 218-29.

2

Satellite Broadcasting: The Regulatory Issues in Europe

Rosemary Hughes

PRINCIPLES OF REGULATION

The historical reasons for the regulation of broadcasting are based on a number of principles, which are interdependent and which have been extended to control satellite broadcasting:

- TV has always been seen as a politically important medium —not least as a cypher for social persuasion
- technically, it has been necessary to control frequencies, to avoid interference with other parts of the radio spectrum
- 'new media', ie cable and satellite, have assumed a special importance for reasons of industrial strategy, and thus must be made to fit into policies on employment, exports, etc
- the original link with telecommunications authorities, for licensing purposes remains in most countries, so that regulation of cable and satellite still falls within the remit of telcommunications organisations/PTTs (Posts, Telegraphs and Telephone companies)
- broadcasting has long been seen as a 'public service' (in some cases, like utilities, such as gas and electricity), which must serve the 'public good', in terms of requirements not to offend public decency and taste, avoid political bias, provide a range of programmes for different sectors of society.

Politicians have long believed that television is a real force for political communication to the public. State control is recognised as the best way of guaranteeing that broadcasting does not become an instrument of narrow sectional interests. This takes a number of forms, from ownership of transmitters to government appointed watchdogs, overseeing the operations of broadcasters, and control is exercised to varying degrees. Thus, while responsibility for the regulation of British broadcasting is devolved to 'independent' bodies such as the IBA or BBC Board of Governors, the Greek broadcasting industry is controlled by the Ministry to the Prime Minister and private interests are rigidly excluded from TV and radio.

The political supervision of broadcasting has traditionally been justified by reference to scarce radio frequencies. Off-air broadcasting uses scarce frequencies in the radio spectrum, which necessarily constrains the number of channels available to the public. Thus, government regulation is necessary to allocate frequencies and prevent interference among different stations. Cable and satellite television, by using broadband technology, offer access to a greatly increased range of channels.

New media expansion is often perceived as a source of new wealth by European governments. In the UK, 'cable and satellite expansion are part of an attempt to revitalise the economy and place UK electronics industries at the forefront of technological development in telecommunications in Europe'.[1] The Thatcher administration also believed that cabling Britain would create new jobs, as well as enhance manufacturing income.

In other countries, a similar emphasis has been laid on the importance of satellite communications for the growth of strategic industry sectors, such as electronics and aerospace. The French governments have invested a great deal of political capital in the new media programmes. The ambitious Plan Cable had as its objective the cabling of France by the end of the century, concentrating on optical fibre networks (such as that in Biarritz) which provide both information and entertainment services. Much kudos was also associated with the joint Eurosatellite manufacturing venture between France and West Germany in the construction of Europe's first DBS satellites — TDF-1 and TV-SAT. Herr Schwarz-Schilling's equally ambitious scheme to cable West Germany at a rate of 1.5 million extra homes per year was also based on the Minister of Post and Telecommunications' belief in the long-term value of interactive services.

The market for satellite services, therefore, has assumed an economic significance as part of a strong technological base for the promotion of economic growth and competitiveness. The 'software' industry is also booming, with an increased demand for good quality, reasonably priced TV programmes to fill the extra hours of broadcasting.

Broadcasting (has been) subsumed in the information technology revolution and in the efforts of European governments individually — and, to a certain extent, collectively — to capture rapidly expanding international markets for information, entertainment and electronics.[2]

A combination of technical and political forces has allowed the control of broadcasting to be shared between public broadcasting institutions and telecommunication authorities. The legitimation for this control was the limited number of airwaves and the need to prevent frequency interference. Thus, the national PTTs became the licensors for radio and television signal transmission and reception.

But the telecommunications administrations' continued involvement in broadcasting first prompted and then underscored the high level of capital investment required to develop transmission networks — from microwave, through cable to satellite. The role of network constructor became self-fulfilling, since only at the level of state funding could such infrastructures be maintained. Private participation in telecommunications, where permitted, was restricted to the provision of service software.

Of all European PTTs, the Deutsche Bundespost plays the most dominant role in the development of cable and satellite television. The Bundespost's commitment to the introduction of ISDN involves long-term plans to extend the 2.5m-homes national cable grid, and the upgrading of coaxial systems with fibre optic. This will enable the delivery of an enormous range of interactive services to homes and businesses in the final phases of ISDN development. At the same time, the German telecommunications organisation has invested DM180m (£60m) in the construction of Europe's first DBS satellite, to be launched in 1987.

In all European countries, the PTTs, as signatories to Intelsat and Eutelsat, hold the monopoly of uplinking, and in some cases of downlinking from these satellites. With the launch of high-power DBS satellites, however, while PTTs may continue to be responsible for the licensing of the receiving equipment (as regards technical approval), there will otherwise be no bar to the downlinking of such signals by individual householders.

A gradual reduction in the PTTs' hold over telecommunications services is also evident in moves towards the deregulation of European telecommunications, inspired by an EEC initiative. The impact of this for satellite TV is investigated more fully in the final section of this chapter.

The concept of broadcasting as a 'public service' underpins the charters of most European broadcasting organisations, embracing 'a commitment to quality, to providing the same level of service to the entire country irrespective of geographical location or social and economic circumstance and to presenting issues impartially and in a balanced manner'.[3]

This model, exemplified by the BBC and copied by many Western European broadcasters, has, however, been traditionally supported by the principle of monopoly. Thus, TV and radio services were vested in a single national broadcasting institution, such as NRK of Norway, SVT of Sweden, RAI of Italy and Elliniki Tileorassis of Greece. Even in the Netherlands, where the broadcasting system is obliged to reflect the traditional 'pillarisation' of Dutch society, the various private broadcasting organisations operate within the strict framework of the Nederlands Omroep Stichting (NOS).

Inevitably, the new competitive media environment threatens this public service model. The legitimacy of the established national broadcasters is being challenged by the irrelevance of frequency scarcity, the national coverage offered by satellite channels and the extensive fragmentation of the audience. Moreover, as video cassette recorders and new channels compete for the viewer's time, the consequent reduction in audience ratings makes the case for politically unpopular licence fee increases even less defensible to governments. The broadcasters feared that their public service obligation could become a straitjacket. The next section shows how the broadcasters are dealing with these changes.

NATIONAL MEDIA POLICY

National media policy has traditionally rested on the principle of maintaining a public service broadcasting system, supported by a monopoly of TV and radio transmissions, if not necessarily on a regional/local level, certainly at a national level. This domination of the airwaves has been supported by regular revenues from public funds, from a licence fee or from a monopoly on national TV advertising.

This monopolistic model is now being challenged in a number of European countries. The impetus towards change has come from a number of different sources:

- politically opposed forces
- widespread dissatisfaction with the existing broadcasting institutions, either due to programming policy or for more ideological reasons
- greater competition afforded by new communications technology providing alternative/additional transmission media for TV programmes
- pressure from commercial circles for an increase in advertising opportunities
- a concomitant liberalisation in telecommunications

In many countries, the conventional swing between Left- and Right-wing governments is reflected in the differing attitudes to the mass media. The current moves to create new broadcasting channels, or encourage new media technology, are related to the political ideology of the party in power.

The present liberalisation of the French broadcasting scene begun by the Socialist Administration in 1984, may seem to have contradicted Mitterand's policy of nationalisation. But, in fact, the authorisation of three new private channels — Canal Plus, La Cinq and TV6 — in 1984/5 by President Mitterrand was rooted in a desire to neutralise the impact of the state TV system falling into the hands of the Right. Crucially, although the concessions would be awarded by the government, their day-to-day operation would be outside the government's influence. The choice of Silvio Berlusconi to head La Cinq was attributed to his Socialist sympathies, although this seems unlikely given his enormous success as an entrepreneur. In this way, politically-opposed forces created the climate for the privatisation of the airwaves, even though the Socialists' advocacy of deregulation runs counter to their expressed ideology.

Another effective spur to change, as in any sector of society, is that based on dissatisfaction with the *status quo*. Thus, the perceived complacency of the BBC in the 1950s led to the introduction of competition in the form of commercial TV, which successfully broke up the Corporation's thirty year monopoly of the airwaves. Its radio programmes had, of course, long competed with Radio Luxembourg, but TV still operated under a Reithian ethos of paternalism — 'worthy', but ultimately dull programmes, which alienated many sections of the audience, with their refusal to recognise the demands of a culturally divided society.

A similar situation pertains in Spain, where RTVE has long been renowned for its obvious support of the Socialist administration. (This was particularly noticeable during the NATO referendum campaign.) After several years of lobbying by the Opposition parties, a law has been introduced in 1987 allowing for the creation of three private national channels. In the short term, the Opposition succeeded in changing the way the Director-General is appointed. Instead of relying on a Cabinet decision, the latest Director-General, Sra Pilar Milo, was elected by a three-fifths majority in parliament.

The new communications technology, of course, is playing a vital role in the expansion of the media scene. Belgium cable TV audiences have long been accustomed to receiving programmes from foreign countries. Now, through satellite dishes, more countries will be able to enjoy the same increase in channel availability. The choice of viewing is no longer limited to the fare offered by national public service broadcasters. This pattern has been described as the 'decentralisation' of TV and, because much of the programming — for UK cable channels at least — is bought 'off-the-shelf' from relatively cheap Australian and American studios, we are witnessing also the 'internationalisation of supply and reception of content'.[4] The latter phenomenon is, of course, not new to British, French, Italian and Greek screens, for instance. But, the inclusion of 'foreign' material is usually controlled either by a government-imposed quota or, as in the UK's case, by voluntary agreement. (Thus, no more than 14% of programmes shown by the BBC, ITV and Channel 4 should be of foreign origin. What is not controlled, of course, is when this restricted amount of programming may be shown!)

The increased supply of programmes from different sources is leading to a fragmentation of the TV audience market. One result of this is to undermine the justification, in terms of audience numbers, for financial support through the licence fee. If more and more viewers turn to cable or satellite channels, then the national broadcasters may be accused of not achieving their public service aim.

Besides calls from the audience for more choice in programming (the level of dissatisfaction with existing channels varies between countries), there has for long been a strong commercial lobby demanding more advertising outlets (especially in the light of foreign channels taking advertising revenue out of the domestic market). Across Europe, though excluding the UK, there is a serious shortage of advertising airtime on national channels. Commercial TV is prohibited in Scandinavia and brand promotion is not permitted in the Netherlands and Belgium, while commercial airtime is heavily restricted in West Germany, Austria, Switzerland and Greece. The increase in advertising slots, while welcomed by advertisers and the commercial channels, is inevitably perceived as a threat to the advertising income of the press. It is not surprising, therefore, that stakes in many new TV channels are held by or are sought by publishers (cf Hersant's 35% share in La Cinq; the majority holding of a number of publishers in SAT1; Robert Maxwell's ownership of UK cable channels and networks, and stake in TF-1; etc).

Cable and satellite television have raised new issues that cross the boundaries of traditional policy communities in tele-communications, industry and broadcasting. The new media offer not only opportunities for an expanded entertainments sector, but also a path to the 'wired society' via interactive broadband cable and small dish business services (VSATs). Technological convergence erodes the boundaries between telecommunications and television, just as telematics combines telecommunications and computing.

Telecommunications authorities compete with broadcasting authorities for control over valuable new outlets and, thus, new revenue streams, hence the current dispute between TDF, the satellite organisation, and the DGT (Direction Generale des Telecommunications), who see TDF-1 as a competitor to the PTT's Telecom-1. Developments in satellite broadcasting cannot be taken in isolation from the telecommunications services, whose transponder capacity they currently share. Hence Eutelsat's distress at British Telecom International's involvement with a rival medium-power satellite carrier, SES/Astra (see final section).

A radical restructuring of national broadcasting environments is, nevertheless, now taking place across Europe, with the release of extra terrestrial frequencies. Figure 2.1 summarises these moves. The advantages of these to governments is that they can be controlled, while, from the programmer and advertiser's point of view, they can provide blanket coverage. Furthermore, the continuing delay in the launch of satellites through Ariane failures and slow cabling of Europe, mean that many new terrestrial

channels will be satisfying the audience desire for more programme
choice before expensive satellite operations are underway.

Figure 2.1: Developments in European Broadcasting, 1985-1990

Austria	a new broadcast channel has been proposed by the state broadcaster, ORF; likely to go ahead without any increase in overall commercial airtime
Belgium	the new private commercial channel, TVi, two-thirds owned by CLT, the Luxembourg broadcaster, will start transmission in the French-speaking region of Wallonia in autumn 1987; a bill has also been passed by the Flanders regional parliament authorising a third Flemish channel
Denmark	a second TV channel, intended to carry the first TV advertising in Denmark, began regional transmissions in June 1987; TV2 is operated by a state-owned body, independent of the existing state broadcaster, Danmarks Radio; it is intended to develop into a national service by October 1988
Finland	a third TV channel, 50% owned by the state-owned broadcaster, YLE (Yleisradio), is also undergoing regional tests; it will eventually be financed by a mixture of advertising and subscriptions
France	considerable upheaval in the media scene over the last two to three years has resulted in the creation of three new private commercial channels - Canal Plus, La Cinq and M6 - and the privatisation of TF1, the most successful state-owned channel; further cession of control in state TV is planned by the right-wing Government
Italy	public broadcasting, in the form of RAI's three channels, continues to battle for audience share with Silvio Berlusconi's three private channels, Italia Uno, Retequattro and La Cinque; efforts to introduce some order into the chaos through a media bill have been continually thwarted by political instability
Netherlands	in an effort to counter the competition for advertising revenue presented by UK satellite channels, a third state-run national channel is scheduled to start in April 1988; it will not itself carry advertising, but will be funded by an extra hour of commercial airtime a week on the first channel, Nederland 1
Norway	opposition to commercial TV remains strong in Norway, despite the decision of its neighbour, Denmark, to introduce a new advertising channel; none of the key issues surrounding the much-debated second channel has yet been resolved
Portugal	the introduction of two new channels has been proposed, with legislation likely in 1988; commercial airtime will increase
Spain	legislation is underway for the creation of three new private channels; much interest has already been shown by Silvio Berlusconi and Robert Maxwell

Sweden	despite the freedom accorded satellite television, the Government remains opposed to commercial TV, which might prevent potential advertising revenue going to foreign channels; Opposition plans refer to a third channel, independent of the Swedish Broadcasting Corporation (SVT) and carrying advertising
Switzerland	the Government's draft bill on broadcasting, published in 1986, marks the first step towards ending the monopoly of the three-channel state broadcasting system, SBC; a private operator will be allowed to use the last of the country's four national TV frequencies
West Germany	in 1986, the DBP released 90 low-power terrestrial frequencies to extend the availability of programmes; these have been shared between private companies and ARD, the state broadcaster; a media treaty by the 11 state governments, concluded in March 1987, established a proper infrastructure for the working relationship of private and public channels and, inter alia, extended commercial airtime for both sectors

If Italy did not set an example in controlled broadcasting deregulation, it certainly led the way in 1975 when the Constitutional Court allowed experimental private networks to continue locally, in barely controlled competition with RAI,the state broadcaster. Other European countries have taken the cue much more hesitantly, and have permitted the 'demonopolisation' of broadcasting institutions within a strictly regulated framework.

Concern about the cultural and, not least, the eventual political impact of changed media patterns has created national measures to counter potential threats to existing public service broadcasting institutions. In some cases, the new laws will serve to support existing broadcasting institutions by allocating them a dominant role in new channels or increasing their advertising revenue potential. For example, the allocation of new local terrestrial frequencies by the West German PTT is being carefully divided between public and private broadcasters. Moreover, in Finland, Oy Yleisradio AB, the state broadcaster, has a majority stake in the new commercial third channel, Oy Kolmostelevisio AB, which must also operate within the terms of YLE's licence.

In addition, the broadcasters are harnessing their experience in programme provision and vast resources to join in the new media contest. Satellite 'omnibus' channels, such as 3SAT and TV5, set up by broadcasters, offer an additional outlet for programme archives and some also invest in satellite channels. Moreover, the principle of 'must-carry', which demands that cable operators

57

Figure 2.2: The involvement of European broadcasters
in satellite TV

Satellite channels

Austria	ORF contributes programming to 3SAT, the German cable channel
Belgium	RTBF, the French-language broadcaster, contributes programming to TV5, the French cultural cable channel
France	TF1, A2 and FR3, the state broadcasters, all contribute programming free of charge to the cable channel, TV5
Germany	ARD provides programmes for its satellite venture, Eins Plus; ZDF contributes programmes to 3SAT, another outlet for archive material
Ireland	RTE has been involved in attempts to revive Europa, the public service consortium channel, instigated by the EBU, which collapsed in 1986
Italy	RAI has also played a dominant role in Europa
Luxembourg	CLT, through its TV station, RTL, supplies programmes to the German-language cable and terrestrial channel, RTL Plus
Portugal	RTP is another key player in the Europa satellite channel
UK	BBC & ITV - provide programmes to Super Channel, a pan-European cable channel, combining broadcasters' archive material with Music Box, a 24-hour pop music programme; all ITV channels, except Thames, also contribute to the channel's operating costs; Thames supplies programmes and investment to Children's Channel; TVS provides programmes and investment to Lifestyle and programmes to Arts Channel; Yorkshire also provides programmes and investment to Lifestyle, and has a 40% stake in Music Box; Granada has invested in Music Box; Central TV has invested in Children's Channel; Tyne Tees provides programmes for Lifestyle; S4C has provided programmes for Children's Channel

Satellite Ventures

Finland	YLE has been allocated a transponder on Tele-X, the Scandinavian DBS satellite
Germany	Both ARD and ZDF have been allotted transponders on TV-SAT, the DBS satellite
Italy	RAI has reserved transponders on Olympus and Sarit, the national high-power satellites
Luxembourg	CLT has options on transponders on TDF-1, the French DBS satellite, on TV-SAT (for RTL Plus) and on SES/Astra, as the Luxembourg national broadcaster
Norway	NRK has been allocated a transponder on Tele-X
Sweden	SVT has been allocated a transponder on Tele-X
UK	Granada and Anglia TV have stakes in BSB, the UK DBS consortium

---

include public service channels in their schedules, guarantees their distribution via new media. This will be further extended for the German broadcasters, ARD and ZDF, and for the Nordic broadcasters, SVT, NRK and YLE, who have already been allotted transponders on their respective national DBS satellites, TV-SAT and Tele-X. Figure 2.2 details the involvement of broadcasters in satellite television.

THE REGULATION OF SATELLITE BROADCASTING

This comprises two areas, which might neatly be summarised as 'hardware' and 'software'.

Technical Regulation

There are two principal components of the technical regulation of satellite services —
- the space segment, determining the configuration of the transmission system
- the ground segment, legislating the reception of satellite signals

The space segment

Historically, the International Telecommunications Union (ITU) has been responsible for controlling each new dimension of radio communications. This is based on the notion that radio frequencies, and in the case of satellite communications, orbital slots also, are limited commodities.

Those areas which fall within the remit of the ITU include:
- allocation of orbital positions and spacing
- regulation of the power of beams and the shape of beam footprints
- frequency planning and allocation

The structure of the ITU is shown in Figure 2.3. The members are representatives of the radio regulatory departments from 161 countries. They meet at World, Regional or Extraordinary Conferences (WARC, RARC, or EARC).

Traditionally, frequencies and orbital placings had been decided on a 'first come, first served' basis. But, DBS satellites seemed to constitute a new, relatively discrete application for satellite technology. The allocation of orbital slots and division of the

Figure 2.3: Structure of the ITU (International Telecommunications Union)

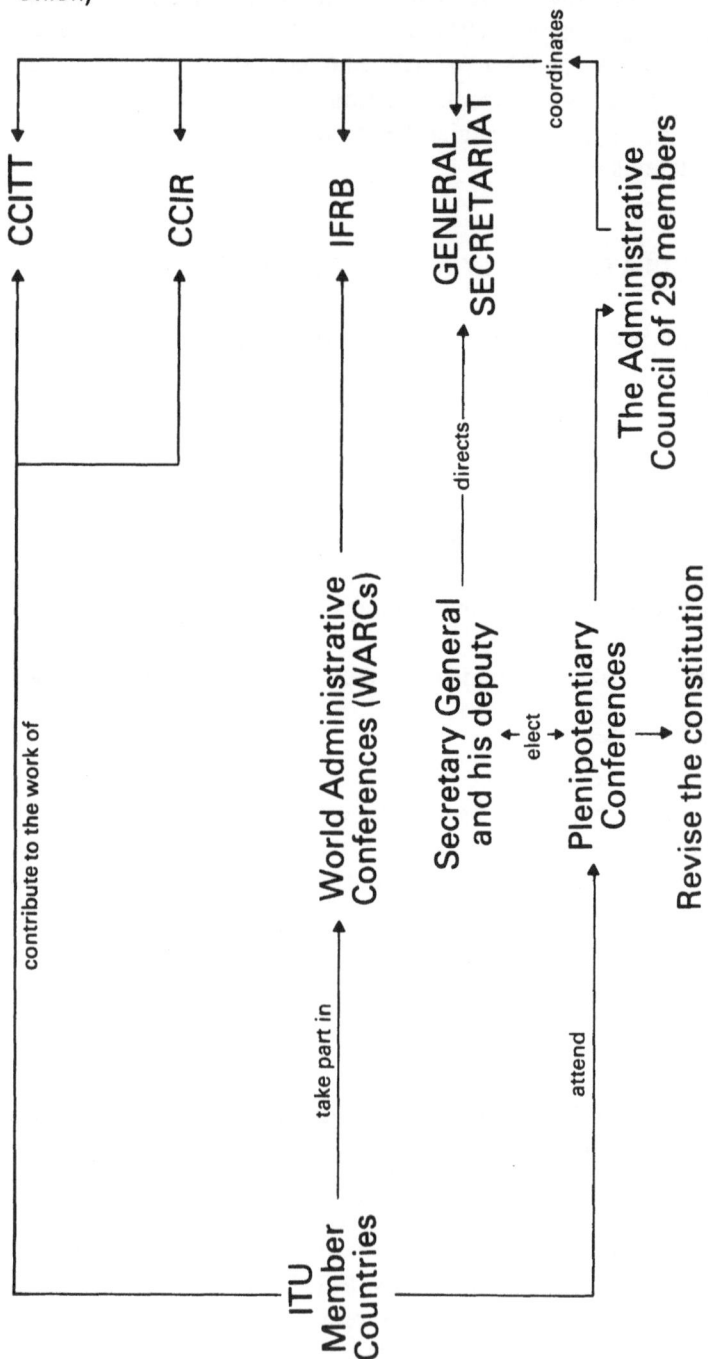

frequency spectrum were made on an *a priori* basis, for the first time, by the 1977 WARC.

However, in the light of successive technological advances, both in satellite transmission systems and reception equipment, the pre-assigned DBS slots now seem inappropriate. Indeed, it could be argued that WARC'77 has effectively 'sterilised' many of the orbital positions and frequency ranges until a further international agreement can be reached. In the meantime, a new generation of medium-power 'birds' has superseded the ambitious plans for DBS satellites begun ten years ago.

A further technical problem for which WARC'77 tried to cater was that of *overspill*. Satellites are, inevitably, no respecters of national boundaries, particularly the larger platforms with a transmission power ten times that of existing telecommunications 'birds' (i.e. 230 watts/channel versus 20 watts/channel). DBS, however, was intended as a *national* broadcasting medium and WARC'77, therefore, laid down requirements for the shape of such satellites' beams and the strength of the signal power in order to minimise overspill as far as possible.

In practice, of course, satellite broadcasters are unlikely to discourage the reception of their programmes over as wide an area as possible. For, in order to maximise advertising revenue potential, international markets will be sought and the increasing sophistication of receiver technology will help the broadcasters in this objective.

Apart from the issues of technical redundancy and commercial exigency, it could be argued that the WARC'77 overspill limitation arrangements contravene the European Convention on Human Rights, Article 10 of which allows for the right of everyone to freedom of expression and thus programmers' freedom to transmit.

The ground segment

As we have already indicated, improvements in receiver performance and moves towards medium-power satellites, will open up the market for cable channels, currently aimed at European cable network headends. Soon, smaller antennae (e.g. 85-90cm diameter), suitable for individual homes, will allow the direct reception of such channels, as well as new DBS programming.

It is important to point out the ITU distinction between *fixed satellite services (FSS)* and *broadcast satellite services (BSS)*, in order to understand the regulatory confusion surrounding new satellite programming caused by rapid technological advances.

FSS — a radiocommunication service between earth stations at specified points when one or more satellites are used; or for connection between one or more earth stations at specified fixed points....

BSS — a radiocommunication service in which signals transmitted by space stations are intended for direct reception by the general public. [ITU Radio Regulations, 1983]

The FSS category was originally devised to accommodate PTT point-to-point telecommunications services, such as telephony, transmitted on low-power satellites. But, the increasing use of FSS satellites for TV distribution to cable networks makes it possible to have a broadcast satellite service also operating on an FSS satellite. Moreover, improvements in receiver technology mean that fixed satellite services are fast becoming the forerunners to DBS through their potential to serve individual homes.

Reception of such signals would be by a limited number of members of the public who had contracted with the programme provider (e.g. cable operators). Transmissions via FSS may be scrambled so that they can be received only with the appropriate decoder, as is the case with Sky Channel. (Unauthorised interception of such broadcasts, and their decryption, is, nevertheless, becoming increasingly common.) A BSS, on the other hand, would be reserved for high-power DBS satellites.

Services carried on communication satellites, such as Eutelsat I and Intelsat V, generally fall outside national *broadcast* regulations, which govern issues of taste and decency, for instance. Their downlinking, however, is subject to the national PTT's monopoly of *telecommunications* services provision, which insists on the licensing of satellite TV reception. As regards DBS, the general understanding of the WARC'77 agreement, and the subsequent resolution by the EEC and Council of Europe, is that DBS is both *broadcasting* and *national* in focus. No PTT licence, therefore, should be required for the individual reception of programmes in the broadcast spectrum.

In the meantime, pressure from different sectors is gradually weakening the PTTs' hold on satellite TV reception. Individual reception of satellite channels from communications satellites has been legalised by most telecommunications administrations, according to varying degrees of restriction. Figure 2.4 summarises the regulatory requirements for FSS TV reception in Europe.

Figure 2.4: Regulatory requirements for communications satellite TV reception in Europe

Austria	dish installations in a cabled area are not allowed; otherwise, the PTT issues licences
Belgium	the PTT issues licences for individual dishes
Denmark	recent liberalisation of satellite TV reception, with licences issued to dish owners by the Teleinspektionen
Finland	there is no prohibition on the reception of satellite signals, neither are there any licensing requirements; dish owners are, however, required to pay an annual fee per channel received to the PTT
France	a one-off free licence is granted by the DGT for domestic installations; although private dish reception is not prohibited, the right of a CATV operator to have exclusivity of satellite signal distribution is yet to be challenged
Germany	recent deregulation of direct satellite reception involves a one-off licence application and a monthly charge to the PTT (DBP)
Ireland	only recently has legislation allowed the downlinking of satellite programmes by cable operators; individual installations are permitted
Luxembourg	a free licence is required from the PTT and all dishes should be installed in accordance with the regulations of the municipality
Netherlands	no licence is required for domestic dishes
Norway	no licence is required for individual installations
Sweden	no licence is required for individual installations
Switzerland	a monthly per channel fee is payable to the PTT for any dish installation
UK	reception of satellite TV services requires a licence from the DTI (3,000 issued up to May, 1987), except for domestic installations where the dish is less than 90cm

Note
There is, as yet, no legislation dealing with satellite dish installations in Italy, Spain, Portugal or Greece.

The regulation of satellite broadcasting coverage and content — European initiatives

The facility of satellite communications for directing messages at large geographical areas emphasises the potential for conflict between countries and, thus, the need for international co-operation. However, there are few internationally binding legal instruments to regulate satellite broadcasting - merely recommendations for the treatment of issues, such as copyright and advertising, such as the Berne Convention (1886), European Agreement on Protection of Television Broadcasts (1960) and the Draft Resolution of the Council of Europe on Advertising (1984). On the whole, national laws continue to form the strongest basis for legislation in this field. The fact that satellite transmissions necessarily cannot be restricted to national boundaries raises areas of dissonance between countries on certain fundamental issues.

> With satellite broadcasting, the European dimension of the new communications technology was even more apparent. Indeed, by comparison, cable TV seemed to provide an 'electronic Maginot-line', enabling governments to continue to control viewing.[5]

One of the main European debates, in fact, concerns whether broadcast satellite signals should be considered to fall within national or European cultural legislation.

National laws on satellite broadcasting, therefore, stand beside European charters. European-wide regulation and promotion of satellite TV are of interest to four principal bodies -

* *European Broadcasting Union*, which co-ordinates the international exchange of programmes between its members and promotes common technical standards
* *Eutelsat*, the satellite carrier whose members are drawn from the European PTTs, and which, with Intelsat, jealously guards a duopoly of satellite transmission capacity across Europe
* *Commission of the European Communities (CEC)*, which wants to create a 'common market' for broadcasting, as exists for other European 'goods', among the twelve member countries
* *Council of Europe*, which was set up with the signing of the Human Rights Convention in 1949, and is primarily interested in social issues, including broadcasting and the mass media in general; unlike the CEC, the Council covers all 22 countries of Western Europe.

The latter two are particularly concerned to rationalise individual nations' broadcasting laws and impose a pan-European structure on programming legislation.

The Commission of the European Communities (CEC) believes that the expansion of services by satellite demands supra-national regulation. In the same way that the Commission encourages the free flow of goods and services across frontiers, it believes that intellectual property, in the form of television programmes, should also be allowed trans-national freedom of movement.

This 'common market' for broadcasting found expression firstly through its Green Paper, *Television without Frontiers* (1984), and subsequently via a *Draft Directive* (1986).

Television without Frontiers had three principal aims:

- to demonstrate the importance of broadcasting for European integration
- to illustrate the significance of the EEC Treaty for those responsible for producing, broadcasting and retransmitting radio and television programmes, and for those receiving such programmes
- to submit for public discussion the Commission's thinking on the approximation of certain aspects of Member States' broadcasting and copyright law before formal proposals were sent to the European Parliament and the Council.[6]

As well as emphasising the cultural advantages of a 'free skies' policy, the Commission was obviously concerned about the commercial implications of restricting information flow. It believed that the protectionism evinced by national broadcasting regulations was futile in the face of advances in media technology.

Lord Cockfield, the Commissioner in charge of television policy, argued that:

> national television companies will face, as a matter of course, free competition from broadcasters established elsewhere in the Community. Only attractive and good quality programmes will retain the audiences now too often held by national monopolies.[7]

Moreover, continued support for such chauvinistic systems would discourage European initiatives and thus prevent the full exploitation of new technology opportunities. The Commission recognised the threat to European manufacturers of satellite technology from the Americans and the Japanese, unless some measure of product standardisation could be agreed on. The continued refusal to agree on a common European video transmission standard for satellites will ultimately fragment an uncertain market and will certainly deter manufacturing commitment to the mass production of decoder chips, on which economies of scale and, hence, low prices are dependent.

The Commission proposed three principal measures:

- co-ordination of specific aspects of Member States' laws regulating radio and television advertising
- limited co-ordination of Member States' copyright laws to ensure that copyright holders' rights to prohibit the simultaneous transmission of programmes coming from other Member States are everywhere replaced by rights to receive fair remuneration
- co-ordination of certain aspects of Member States' laws regulating broadcasting in the interests of fair play ('right of reply') and of protecting children and young persons.

National broadcasters would, however, be free to set their guidelines within this framework. So long as a broadcast met the basic EEC norms, it could not be stopped from crossing frontiers. Thus, the Netherlands and Belgium would not be able to continue banning advertising from foreign satellite channels re-transmitted via cable.

The Green Paper quickly roused strong feelings among broadcasters and politicians, largely critical of its suggestions. Nevertheless, a Draft Directive was published in March 1986, which extended and clarified the Commission's commitment to overcome what it saw as the major stumbling blocks in the free circulation of programming — divergent rules on advertising, copyright problems and programme quotas. In attempting to 'sweep away the entangled underwood of national regulatory obstacles to admitting broadcasts from other Member States'[8] the CEC made the following precise recommendations:

- a maximum of 15% airtime to be allocated to advertising
- a minimum of 30% of non-news programming to be produced within the EEC, rising to 60% over three years
- independent production to account for 5% of programmes, rising to 10% after three years
- the advertising of tobacco and tobacco products to be banned, and the advertising of alcohol to be restricted
- the right of copyright holders to prohibit the simultaneous transmission of programmes coming from other member states to be removed; instead, copyright holders should have the right to receive fair remuneration
- the MAC-packet family of standards for satellite transmissions, including both C-MAC and D2-MAC to be accepted.

Protests of equal force greeted the publication of the revisions, from those who insisted that programmes could not be treated like any ordinary economic good. Of particular concern to television companies was the suggestion for programme quotas, as the established public service broadcasters saw their programme production monopolies under threat, and satellite channels, dependent on cheap, imported programming, were equally alarmed.

As Alastair Tempest of the European Advertising Tripartite points out:

> Any goodwill the Commission may have had from transnational broadcasters, or with the new satellite operators, is likely to evaporate when faced with such precise programming requirements. These quotas will also strengthen the publc broadcasters' belief that the EEC has over-reached its competence, and reinforce their opposition in principle to all the proposals. Even if they pass into EEC law, it is difficult to imagine that the proposals will provide some sort of guarantee of quality.[9]

Advertising is no less a touchy subject, as the rules restricting both its airtime and content vary widely across Europe (see above). Satellite channels are afraid that limits on advertising will restrict potential revenue. The limitation of advertising time is, however, accepted by a majority of broadcasters. But, there is no agreement on block advertising and many feel that decisions on the length and intervals of the spots should be left to broadcasting and the dictates of the market.

The EEC's proposal to 'substitute an equitable compensation' for the copyright holder's power to prohibit simultaneous cross-border transmissions, in fact, runs counter to the ruling of the European Court in the case of Coditel. This held that the 'exhaustion of rights' (i.e. the freedom to undertake parallel import into any EEC country) does not apply to performing rights and intellectual property. Half of the twelve countries reject the compulsory copyright licences. As a compromise, a European arbitration body has been suggested.

A 'parallel strategy'

The Directive has since become entangled in 'a kind of parallel strategy pursued by some member states who would prefer to see the Council of Europe get involved in the act'.[10] At a confernce on mass communications policy held in Vienna in December 1986, a declaration supported the Council of Europe as 'the most suitable institution in Europe for the elaboration and further development of an appropriate framework for transfrontier broadcasting'. The Council was urged to draw up a legally binding convention 'to provide the appropriate means for preventing or solving possible conflicts caused by the transfrontier development of the mass media...'.[11]

Support for a Council of Europe convention has come mainly from terrestrial broadcasters and the EBU. The UK government, for instance, is openly hostile to the Commission's plans, questioning the competence of the Commission in dealing with cultural issues.

The basis of any Council convention would be its recommendations on TV advertising and the use of satellite capacity adopted in 1984 and outlined below:

- advertisers should comply with the law applicable in one country
- advertising should be clearly separated from programmes, and should not influence programme content in any way
- programme standards, related to taste, decency, news impartiality, violence, racial hatred and the protection of children should be adhered to
- the lessor of the satellite capacity (usually the PTT) has to be satisfied that there is a relevant law that applies to the programme service (in case such services would not be covered)

- the 'transparency' rule says that where capacity is leased, the lessor should make publicly available details of the lessee and the purpose
- voluntary copyright agreements should, as far as possible, be reached between distributors and programme providers.

It is obvious from this checklist that a Council of Europe convention would quietly ignore the issue of free circulation of programming. Moreover, Council conventions are noted for their flexibility of approach. Rules are drafted by member states, rather than by a separate bureaucracy in Brussels. 'In other words, a Council of Europe approach calls for a consensus about how far-reaching the rules should be. A community directive, on the other hand, can be passed by a majority. As soon as it is adopted, a definite time limit is set (two years) during which national legislations have to comply with it. The Council of Europe procedure leaves it to the member states to act on ratifying a given convention with no precise and compulsory schedule.'[12] Historically, council recommendations have proved to be sogeneral that few objections can be found to prevent their adoption by most countries. 'In other words, they don't change anything. They just provide a formula which accommodates everybody.'[13]

The Commission says that it welcomes the move by the Council, but claims that the draft directive will not be undermined. The Commission argues that ratification of the convention could be a slow process, and says that, in any case, it will be difficult for a convention to ensure the free circulation of programmes. It says that the EEC has to go further than the Council, because it has to break down the internal barriers to trade within the community by the end of 1992.

Résumé

If the Commission's controversial proposals have had any effect, it has been to bring to the fore some of the long-term implications of satellite broadcasting and to focus the minds of broadcasters and governments alike on the main issues associated with transborder television. In attracting such antipathy towards its recommendations, the Commission has opened up national media policy to a wider sphere of discussion. It has brought out into the open the national differences in media policy and made an attempt to resolve some of these.

The Commission's attempt to at least slow down the progressive 'Coca-cola-nisation' of the airwaves, by promoting EEC-originated programmes, has, however, ended up alienating the European organisations it was trying to help. The European programming industry is experiencing spiralling costs, which makes domestic production increasingly expensive. The situation will, of course, be exacerbated as national broadcasters have to fill the increased broadcasting hours. Moreover, experience from Italy and France has shown how European prices for programme acquisition rise markedly once competition is introduced into the marketplace. These countries and the UK are among the biggest importers of programming from outside the community, since American and Australian programmes, bought 'off-the-shelf', allow the broadcasters to fill their schedules relatively cheaply and, furthermore, are popular with audiences.

While advertising restrictions can have serious implications for a pan-European advertising-supported satellite channel, these channels must be encouraged by the moves in many European countries to introduce commercial TV. For, although the new terrestrial channels will mean more competition, it must mean also the relaxation of some of the present awkward bans which prevent specific product promotion.

As regards copyright, while audience figures remain low, minor infringements of release dates have gone unnoticed. This has already begun to change with the advent of pan-European audiences. The launch of Super Channel, for instance, in 13 European countries was seriously embarrassed by the failure of an agreement over copyright and performance fees, which led to the withdrawal of programmes from the schedule at the last minute.

Cheap programming is, of course, vital in the early days of a satellite channel's development, where production costs outweigh revenue and audiences are low. With the growth in Direct-To-Home satellite transmission, programme buyers have become increasingly anxious to negotiate contracts which embrace rights of exploitation over all media, and distribution beyond national boundaries. During a period of enormous expansion in outlets, the evaluation of the potential revenue sources by regulators and copyright holders is almost impossible. In the meantime, the issue of copyright, like transmission rights and advertising permission, is settled on a piecemeal basis, and an EEC solution would certainly force a radical change in this practice.

But before a European solution can be found, either in the CEC Directive or the Council of Europe Convention, the broadcasting landscapes of many countries will undergo radical alterations. Ironically, the true liberalisation of the airwaves is being conducted primarily via terrestrial frequencies and less through the satellite transmission of programmes, as reception equipment remains prohibitively expensive for most consumers and national cabling programmes continue their slow pace. The authorisation of new national channels and release of more commercial airtime to advertisers demonstrate not a new 'deregulation' of television, but a 're-nationalisation of broadcasting policies', as the release of new terrestrial frequencies 'enables national governments to stay in control of developments and to dictate their terms according to national priorities'.[14] Not much hope for a common European approach here.

Where the EEC might more fruitfully impose a common policy is in the choice of technical standard for transmission of satellite programmes — B-MAC, C-MAC or D2-MAC. Here lack of consensus among European manufacturers leaves the way open for an influx of competitively priced receiving equipment from the Far East and America. This is a 'trade' issue on which the Commission could justifiably spend some time in negotiating agreement across national boundaries.

THE IMPACT OF TELECOMMUNICATIONS DEREGULATION FOR SATELLITE BROADCASTING

The PTTs' control of satellite television is helped by their common membership of Eutelsat. This is a European satellite organisation, which is responsible for the design, construction, establishment and operation of a telecommunications satellite system for the international public service provided by the PTTs. The Eutelsat I satellite system was originally conceived primarily for the transmission of telephony and high speed data services. Only the EBU employed the satellite for the occasional interchange of entertainment specials and news, among the Western European national TV networks.

But, in 1982, a British cable channel, frustrated with the slow growth of UK cable networks, began using the experimental OTS-2 satellite to transmit to European cable homes. Sky Channel's innovative use of transponder capacity was gradually adopted by other cable channels, to reach a new pan-European market, and Eutelsat I became a low-power distributor of TV programmes.

In the current climate of deregulation, Eutelsat defends vigorously its duopoly position on trans-European communications satellites, alongside Intelsat. The signatories (PTTs) are pledged, according to Article 16 of the Treaty not to authorise any systems which threaten the viability of the organisation. Yet, British Telecom, Eutelsat's joint largest shareholder with France (16.35%), is in the throes of undermining the solidarity of the European PTTs, by joining forces with a rival private satellite venture, Luxembourg's SES/Astra.

The medium-power satellite, to be launched in 1988, will have 16 transponders, 11 of which have been reserved by British Telecom International for British programmers. This will allow BTI to 'co-locate' the cable channels, currently divided between Eutelsat I F-1 and Intelsat V F-11, which will allow their reception via a single dish. Although BTI has also taken an option on eight transponders on the first of the medium-power Eutelsat II series, this will be launched after SES/Astra. Furthermore, considerable damage has already been done, if only in terms of prestige lost, to Eutelsat's monopolistic status.

71

British Telecom found itself trying to fulfil two obligations at once — as a signatory to Eutelsat and the pressure to answer demand for more transponder capacity from its customers. The PTT's putative agreement to pay cash compensation to the satellite organisation may go some way towards alleviating its unpopularity with other telecommunication organisations, but they have been 'left in no doubt that the fight is on with BT, which seems determined to take the lion's share of the transponder lease and uplink business and will now be actively marketing in their territories'.[15] As one programmer pointed out, 'the bad news is that any programmer hoping that the dawn of Astra would mean a real free market in transponders must be suspicious when one existing supplier moves into a virtual monopoly position'.[16]

British Telecom's deal with a private satellite company exemplifies the threat to the PTTs' position as sole national telecommunications carriers and one telecommunications organisation's response to this challenge. While telecommunications reform is being debated in almost every European country, only Britain has so far acted. The 1984 Telecommunications Act paved the way for the privatisation of British Telecom. A year later, 51% of the Government's stake was sold to private investors. The UK is now not far behind what has been achieved in the US, since the break up of the Bell Companies in the early '80s. Deregulation is only at the planning stage, however, in the rest of Europe.

Pressure to liberalise the telecommunications markets has come not only from growing demand for a wider variety of services and reduced tariffs, but also from EEC initiatives designed to overcome the technological gap which separates the continent from its main (US) competitors. In some areas of high technology, such as fibre optic and communication satellites, Europe is several years behind in development and experience.

One reason for this has been a lack of standardisation across Europe, as the establishment of national technical norms has appeared a useful means of erecting walls against foreign competition (cf MAC debate). In an effort to counteract this trend, the EEC instigated the ESPRIT and RACE programmes, which encouraged collaboration in Research and Development.

The latest move by the EEC Council of Ministers has been to draw up a Green Paper, providing a framework for the European-wide liberalisation of telecommunications. The aim is to stimulate new telecommunications services and create a large unified market for equipment manufacturers in the Community. Without a common approach to deregulation, the objective of a large homogeneous telecommunications market 'would be seriously jeopardised and a great deal of fragmentation inevitable if the member states were to legislate without previous agreement.'[17]

The main proposals include:

- complete liberalisation of the supply of value added services and all terminal equipment
- agreement on standards, frequencies and tariff principles to promote competition across the Community
- the creation of a European Telecommunications Standards Institute to encourage common standards
- an end to telecommunications authorities' regulatory powers, to stop them thwarting competition
- a ban on cross-subsidies in any areas open to competition
- liberalisation of some satellite communication services

The telecommunications industry has traditionally been governed by a rigid institutional structure, which perpetuates the dominant role of national PTTs. Some administrations will, therefore, cede power to private operations (such as SES/Astra or the Irish Atlantic Satellites concern) with reluctance. In the satellite market, satellite dish manufacturers and programmers, anxious to reach the European audience, have been thwarted by the strict telecommunications laws, which stated that only PTTs could 'downlink' communication satellite (FSS) signals. But policies aimed at the liberalisation of state telecommunications and broadcasting monopolies are providing private operators with a growing space in which to maneouvre. By the end of 1985, for instance, most national governments had legalised individual reception of communication satellite signals (although the size of the dish required to pick up these signals (1.2-1.8m) is more amenable to SMATV systems than private householders).

This has partly been due to a reassessment by some national governments of the public monopoly assumptions which support public telecommunication and broadcasting administrations, as right-wing parties, for instance, encourage market-led expansion of the economy. It has also been predicated on the need to encourage the domestic market for satellite dishes for the sake of indigenous equipment manufacturers, particularly in those countries which will launch their own DBS satellites.

Nevertheless, the hard-won power of national broadcasters and telecommunication administrations will not be easily diminished. And ironically, it is the current trend towards privatisation which serves to support the status quo, by extending regulation to set standards, control operating conditions and supervise the boundaries between the public and private sectors.

At the end of the current phase of policy-making and new media development, there is likely to be a good deal more regulation than before, even though there will be more media and more freedom of operation in several respects.[18]

REFERENCES

1. Veljanovski, C.G. and Bishop, W.D. *Choice By Cable*, Institute of Economic Affairs, 1983

2. Dyson, K. 'The Politics of Cable and Satellite Broadcasting: Some West European Comparisons', *West European Politics*, April 1985.

3. Hollins, T. *Beyond Broadcasting*, BRU, 1984.

4. McQuail, D. and Siune, K. *New Media Politics*, Sage, 1986, p.201.

5. Dyson, K. 'The Politics of Cable and Satellite Broadcasting', April 1985.

6. Commission of the European Communities, *Television Without Frontiers*, 1984, p.1.

7. Lord Cockfield quoted in *InterMedia*, November 1986.

8. EEC Press Notice, March 1986.

9. Tempest, A. quoted in *Cable and Satellite Europe*, May 1986.

10. Mario Hirsch, *Cable and Satellite Europe*, May 1987.

11. quoted in *InterMedia*, January 1987.

12. Mario Hirsch, *Cable and Satellite Europe*, May 1987.

13. quoted in *New Media Markets*, 7 January 1987.

14. Mario Hirsch, *Cable and Satellite Europe*, May 1987.

15. *Cable and Satellite Europe*, April 1987.

16. *Cable and Satellite Europe*, April 1987.

17. Karl-Heinz Narjes, Commission Vice-President, quoted in the *Financial Times*, 12 June 1987.

18. McQuail, D. and Siune, K. 'A New Media Order'? *New Media Politics*, Sage, 1986.

3

The Nordic Countries in the Age of Satellite Broadcasting

Lennart Weibull and Ronny Severinsson

INTRODUCTION

On December 9th 1986 the Swedish delegation at the Council of Europe Meeting in Vienna expressed the following view on satellite broadcasting in Europe:

> Although it would be too simplistic to divide Europe into transnational broadcasting countries and transnational receiving countries, it is nevertheless true that, in general, the larger countries, with their more extensive financial and production bases, have been the first to take advantage of the new technological developments and transmit new programme services to their neighbours. The smaller countries...have been importers rather than exporters of the new services.[1]

This statement from the Vienna meeting on possible recommendations for European regulation of transnational broadcasting is a good illustration for not only the Swedish position during the first decade of the satellite broadcasting age but also the position of its Nordic neighbour countries. There has been a fear that the new satellite broadcasting system will mean a cultural domination of the small countries on the periphery of Europe.

The problem of being mainly a target in the international communication process is not at all a new one and certainly not only connected with satellite broadcasting. In fact there is an extensive research tradition in this field, usually labelled cultural domination.[2] It has been shown that small countries, mostly countries of the so called Third World, are culturally dominated by the industrialised world by means of mass communication media.

Empirical evidence has been given in terms of television or video import figures. Figures on the television programme flow within Europe clearly show the smallest countries as the main importers: the top six import countries in Western Europe include all four Nordic countries (no data are available for Iceland).[3]

It would, however, be misleading to conclude that the Nordic countries are victims of a cultural imperialism in general. What the figures refer to is a high television programme import in comparison with other Western European countries, i.e. a dependence on software from abroad. If we take hardware production as another indicator of mass communication independence we find that the Nordic countries are characterised by a huge export, e.g. from Ericsson, Swedish Philips and Salora.[4] As a high-tech region the Nordic countries are interested in exploring the possibility of transnational broadcasting via satellite. They wish to participate in the international market in terms of hardware.

In this chapter, both the cultural and the industrial perspective of transnational satellite broadcasting will be explored from a Nordic point of view. The main issue is how the Nordic area has acted and reacted to the development of satellite broadcasting and, in particular, its effects on the national public service broadcasting systems. Four specific areas will be looked at. Firstly, Nordic efforts to create a Nordic satellite system. Secondly, Nordic policy in relation to satellite broadcasting from abroad. Thirdly, the public's reaction to transnational broadcasting by satellite. Fourthly, and finally, the impact of transnational broadcasting on the Nordic national public service broadcasting systems. By way of background, these broadcasting systems will be examined in the introductory section below.

This overview tries to encompass Denmark, Finland, Norway and Sweden, whereas Iceland is mostly excluded owing to lack of material. The main focus is on Sweden, especially in respect of the political debate and its consequences. This chapter is based on commission reports, media debate and ongoing research.[5] As an analysis of a period of change, it will stress the general trends and social conditions for the actions and reactions of the Nordic countries, especially the relation between cultural and industrial perspectives in the development of satellite broadcasting.

THE NORDIC RADIO AND TV SYSTEMS

Traditionally, the Nordic countries have had a broad base of shared ideas and practices. In all countries the broadcasting system is based on the public service principle. Originally, only one public service company had the sole right of broadcasting. This principle dates back to the radio systems of the 1920s. Later it was also applied to television. The model of broadcasting which was usually copied was that of the BBC with the public funding the broadcasting organisation through a licence fee. But the existence of public service broadcasting organisations in all the Nordic countries does not mean that they had all adopted the same organisational forms. In fact, each country chose its own formats.[6]

The Swedish broadcasting system, for example, was set out in the Radio Act and in an Enabling Agreement between the broadcasting organisations and the state. In the latter, the general objectives of the broadcasting organisations are set out in terms of their overall cultural responsibility as well as their requirement to observe impartiality and objectivity in their programming. The systems in Denmark and Norway are similar to that of Sweden, although the regulations are not as detailed as the Swedish ones. In contrast, Finland has permitted the introduction of commercial TV. Its programmes, produced by an independent television company, are however included in the schedules of the Finnish public service broadcasting organisation. There are other differences between the nations. Finland and Sweden have two television channels, whereas Denmark and Norway only have one. In Sweden, the distribution of radio and television is handled by the Telecommunications Administration; in Finland, the public service broadcasting company also owns its distribution system.[7]

Until the beginning of the 1960s the public service principle was generally accepted in the Nordic countries. The increasing importance of television, however, meant a growing criticism of the sole right rule, granting a monopoly of broadcasting to the public service company. In Sweden the discussion became intense when a second TV channel was proposed. Organised interests were formed, one consisting of business and industry interested in organising an independent channel and the other a loose alliance of those wishing to keep the broadcasting company as a public utility. In the latter group several educational and cultural organisations were active. In the debate the Swedish Broadcasting Corporation, the public service company, entered this 'cultural alliance'.[8]

The confrontation between the 'commercial' and the 'cultural' alliance of the late sixties resulted in a victory for the latter. The second TV channel in Sweden was organised in roughly the same way as the first one and became part of the public service company. A similar solution came out of the corresponding Finnish debate. In Denmark and Norway, the two Nordic countries with only one TV channel each, the debates have been even more intense, mostly because of the demand for two national TV channels, which were regarded as too expensive within the existing public service framework. However, the debates of the late sixties were also the start for a more general discussion, which made the radio and television systems in the Nordic countries a burning political issue. Thus, in all the Nordic countries the national public service systems had lost some of their traditional legitimacy. In the political debate the broadcasting issue turned out to be a partisan (left-right) issue, the Social Democrats being the defenders of the public service, whereas the 'bourgeois' parties generally were in favour of a market oriented approach to broadcasting.

During the beginning of the 1970s the national broadcasting systems of the Nordic countries met another challenge — the desire for local radio and television. It seems reasonable to regard even these demands as a reflection of the criticism of the sole right principle as it applied to the national broadcasting companies. But, equally important, this criticism had some of its roots in a general mistrust of national party politics. At the end of the 1970s, partly as a response to this criticism, the principle of local broadcasting had been accepted in all Nordic countries. But the organisational forms were mainly chosen in accordance with the public service principle.[9]

In general then, it can be concluded that the public service idea has been very strong in all the Nordic countries. During the 1960s and the 1970s the principle of a national public service with a sole right of national broadcasting has been eroded. It has been challenged by both the idea of market oriented broadcasting and the introduction of a local broadcasting system. The change in the national public service has a political history, where the principle of control — state or market — has been the crucial point. These conflicting principles have been a major theme in the debate between the 'commercial' and the 'cultural' alliances. The main argument for public control is the importance of independent, non-commercial broadcasting for the national culture, whereas the proponents for the market principle stress the importance of the audience's free choice. Until the end of the seventies, in spite of the political debate, the principle of public control was the official policy of all the Nordic countries. But during this period a new dimension of broadcasting emerged: satellite broadcasting. The central issue is its likely impact on the nature of national broadcasting in the Nordic countries.

THE NORDSAT PROJECT — THE CULTURAL ERA

The homogeneity of the Nordic broadcasting systems, both in terms of principles and in organisational structure, has brought about considerable exchange and cooperation between the public service companies. In 1958 the so-called 'Nordvision' cooperation was initiated. In 1969 it was described as 'an important part of the Nordic cultural cooperation'. In practice Nordvision means both a coordination of planning within the Nordic area and a regular exchange of programme productions. In 1971 more than 550 hours of programming were exchanged and in 1978 that amount had increased to more than 950 hours with the main categories being documentaries, children's programmes, theatre and drama. On the other hand, the total Nordic TV exchange does not account for more than five per cent of the Swedish Television content.[10]

The idea of the Nordic TV exchange as part of a cultural cooperation is central to Nordvision. This exchange has also been considered very important by Nordic governments. In the beginning of the 1970s the Nordic Ministers of Education and Culture, within the Nordic Council of Ministers, initiated a commission to stimulate further cooperation. The report, 'TV Across Borders', called for an increase in programme exchanges. There were several ideas about how to promote this. In the ensuing discussion Sweden pointed out the possibility of an exchange via satellite. In 1975, another commission was thus appointed. Its task was to conduct a preliminary review of the possibilities of a satellite exchange. The review was discussed in the Nordic Council, a political body representing the parliaments in the Nordic countries. The Council reacted in favour of the general ideas presented and the Nordic Council of Ministers decided to carry out a more detailed analysis, which could be the basis of a concrete decision on a Nordic satellite exchange system — the NORDSAT.[11]

The new commission had a huge task. All political, judicial, technical, financial and cultural aspects of a Nordic satellite system were to be investigated. Its final report, presented in 1979, was written by several expert groups.[12] The main conclusion was that a Nordic TV exchange via a DBS satellite, under the terms of the ITU decision of 1977 (WARC 1977), was realistic. The commission pointed out five main motives for a full scale Nordic TV exchange:

— a desire to promote an increased cultural cooperation among the Nordic countries
— the importance of Nordic broadcasting cooperation in a situation where satellite broadcasting from non-Nordic countries becomes a common phenomenon
— an increased freedom of choice of TV programmes for citizens of the Nordic Countries
— a contribution to better understanding of the Nordic languages
— an improved cultural situation for linguistic and ethnic minorities within the Nordic region.[13]

These motives, mainly stressing the Nordic cultural bonds, are elaborations of the themes used by all commissions working with problems of Nordic TV cooperation. The contribution of the 1979 commission is the idea of NORDSAT as a bulwark against what were seen as the negative influences of Continental satellite systems. The report is very optimistic: a DBS satellite is regarded as a realistic way to realise the Nordic cultural goals. Both technical and financial problems are not seen as important obstacles to a NORDSAT system. Some legal problems, however, are anticipated. The report concludes that it will probably take six years from a Nordic decision to the start of experimental transmission and eight years to the start of the regular operation of the NORDSAT.[14]

In late 1979 the NORDSAT report was sent out for comment to agencies and organisations in the Nordic countries. The reactions were mixed. Generally, two tendencies could be observed. Firstly, there were some important differences between the Nordic countries. Secondly, within each country the arguments tended to follow the left-right dimension of the domestic debate on media policy. It is evident that the Norwegian and Danish attitudes were generally more positive than the Finnish, with Sweden being in the middle. In Denmark and Norway NORDSAT was perceived as a way of getting more than one TV channel at a reasonable cost. But the most striking tendency in the reactions to the NORDSAT proposal was that within all Nordic countries, the cultural organisations and the public service radio and TV companies were the most critical, whereas Pan-Nordic organisations and industrial groups showed the greatest interest.[15]

Rejections of the NORDSAT proposals were usually made on cultural grounds. There was a fear that the NORDSAT idea would mean a threat to the national culture. Although the Nordic concept and the idea of 'bulwark function' against Continental satellites were accepted in principle it was argued that a full scale exchange would, in fact, mean more of a non-Nordic input of TV programmes because all Nordic TV channels are largely dependent on foreign import. But there was also considerable criticism of the high projected cost, and a suspicion that the actual cost would exceed the estimate. It was also felt that more effective Nordic cooperation in the cultural field could be brought about if the money were used in other ways. The public service companies feared that some of the money for NORDSAT would be taken from the budgets of the national TV systems. In fact, all public service companies, except the Norwegian one, rejected the NORDSAT proposal.[16]

But the reactions to the NORDSAT proposal also reflected another tendency, an interest in launching the project for industrial reasons. For example, it is interesting that the Swedish Confederation of Trade Unions (LO) rejected the NORDSAT idea for cultural reasons, whereas the biggest member union, the Metal Workers' Organisation advocated it because of its importance for the Swedish technological industry. In the latter comments the increased freedom of choice was more often stressed than the interest in Nordic cultural cooperation. The positive reactions from the industrial interests, however, were too weak to play any significant role in the outcome. The Nordic Council of Ministers decided to postpone the project but not to give up the idea as such. In 1982, it was decided to continue with the planning but, at the same time, to consider using other satellites as a way of getting the Nordic television exchange programme started. By this time, however, Denmark had left the NORDSAT project and had begun work on a national cable system (see section 5 below).[17]

To summarise, the NORDSAT idea of the late seventies was mainly regarded as a cultural issue on the Nordic level. It was primarily motivated 'on idealistic grounds and not attuned to economic realities',[18] even though some countries saw it as a way of getting a second TV channel. Originally a software idea concerned with Nordic TV exchange, it encountered too many economic and technical problems in the hardware field. In this area there was not sufficient Nordic unity, or at least no readiness for a far-reaching decision on the Nordic level. The national interests loomed too large and the industrial policies were too different.

THE TELE-X IDEA — THE INDUSTRIAL ERA

In November 1981, the biggest Swedish morning paper contained an advertisement from the National Telecommunications Administration (in the following also called the Swedish Telecom) offering new jobs in a satellite project called Tele-X. Tele-X was an experimental system, planned in cooperation with the Swedish Space Corporation (a state owned company for satellite development), the industry, other telecommunications administrations and international space agencies.[19]

Whereas NORDSAT was known through the public debate, Tele-X was not. The Tele-X project also differed in many other ways from the NORDSAT. Tele-X was a Swedish initiative which arose from an interest in industrial development. The project had started back in the sixties, and in the early seventies a national programme of space research was begun. In 1979 the scope of the programme was extended and was considered a national priority, the motive being to give the Swedish space and electronics industry an opportunity to develop their strength in this field. The first satellite to be developed within the space programme was the Viking (launched in February 1986), designed for research in the ionosphere. The second satellite project was the Tele-X, a DBS, mainly intended for TV and telecommunications.[20]

Whilst NORDSAT aroused a lot of political controversy both among and within Nordic countries, the Tele-X project developed with almost no political debate, seemingly being outside the realm of politics.[21] The Swedish parliamentary decision of 1979 was almost unanimous. It is clear that in the 1980s, such industrial considerations have had a growing impact on the development of telecommunications. Indeed, these considerations best explain the different strategies adopted by individual Nordic countries. In a Swedish government bill of 1983 the Minister For Industry (Social Democrat) stressed the importance of a satellite system for the development of Swedish industry. The Danish lack of interest in a satellite system and concentration on a national cable network was announced in terms of industrial policy. The Danish Minister of Communication (Conservative) stated in 1983 that the new cable network would give Danish companies export advantages.[22] In

Norway the 1982 government commission on cable TV concluded that questions of industrial policy were of such an importance that they should be the object of a special commission. This industrial 'imperative' behind satellite policy is certainly not unique to the Nordic countries. But it is interesting to contrast the argument with the solemnly pronounced cultural motives behind the NORDSAT idea.[23]

The Swedish initiative did not mean that the idea of Nordic cooperation in satellite broadcasting was completely abandoned. At an early stage, Sweden invited the other Nordic countries to join the new DBS project. Norway and Finland accepted the idea, but Denmark, for the reasons mentioned above, decided not to participate. Finland and Norway were somewhat reluctant and decided on a minor economic contribution — 3 and 15 per cent of the cost respectively, the difference reflecting the perceived advantages of Tele-X. Thus Sweden will pay for 82 per cent of the Tele-X project.

Tele-X started essentially as a hardware project. The intention was primarily to launch a satellite, rather than serving specified communication needs, although some ideas of TV transmissions had been discussed. In spite of the original hopes, most of the Tele-X is built by West German and French companies, but Nordic industry has contributed some essential parts. Moreover, the home electronics industry, which is quite important in Sweden is considered to benefit from the launching of a DBS satellite. According to the first plans laid out in 1979 the Tele-X was to have been in orbit by 1986, launched by the Ariane rocket. The problems of Ariane have meant that Tele-X has had to be postponed. At the time of writing it is anticipated that the Tele-X will be launched in the middle of 1988.

The postponement of the Tele-X has not been considered entirely negative. In fact, some groups regarded it as a necessity, because of the fact that a definite decision on how to use the Tele-X after the first experimental phase has not yet been made. The Tele-X is planned for both TV and telecommunications services. For TV a three years experimental period of Nordic programming is included in the project. The financing of TV in the long run and the economically acceptable usage of the telecommunication function is still uncertain. The reason for the delay on the software decision is that the original plans of the Tele-X did not contain detailed studies of market demands. The few analyses made have been widely criticised by the potential users for being unrealistic.

A critic has characterised this as the principle of the primacy of hardware over software.[24] In late 1986 the Nordic Council of Ministers commissioned a further study of the market for the services, e.g. commercial TV channels, pay-TV and telecommunications services. The result of this effort is expected to be meagre — there are too many uncertain factors in the total media environment, including the development of cable systems and the introduction of Continental DBS systems. Given these circumstances it is expected that the Tele-X will be mainly used for Nordic television exchange.

It is worth noting that the idea of the Tele-X project has revived interest in the old NORDSAT idea. A Nordic television exchange could, in fact, be achieved via Tele-X, although this would be in a more limited form than the original plans. In November 1985, the Nordic Council of Ministers, representing the Ministers for Industry, Communication and Culture from Finland, Norway and Sweden, decided that there should be two Nordic TV channels available via Tele-X for an experimental period. Programming would consist of material drawn from the regular output of the national broadcasting services. Such programming would not involve extra expenditure. The public service broadcasting organisations with the inclusion of Danish Radio and TV were then commissioned to draw up proposals for such an exchange. To date, all proposals have been rejected on economic grounds. As a further complication, in early 1987 the Norwegian and Swedish public service broadcasters proposed a four channel system. This proposal has meant a new debate since it would involve redesigning the satellite and so delaying its launch.[25]

A special problem for Tele-X has been the anticipated competition with other similar projects under way. For example, the Telecommunications Administrations in Norway and Sweden have not been very interested in taking responsibility for the operation of a Nordic DBS because of their activity in developing a Nordic cable network for telecommunications services. Besides, they participate in the exchange systems of Intelsat. Within the Nordic area, the Telecommunications Administrations have in fact proposed an exchange programme for television broadcasts via communication satellites linking cable networks together. This programme, called the NORDPROG would use transponders on Intelsat and Eutelsat satellites for the transmission of TV signals. Such a system is already used for transmitting Swedish TV to cable systems in Western Norway (via Intelsat) and Norwegian TV to the drilling platforms in the North Sea and to the island of Spitzbergen (via Eutelsat). The transmissions are financed by user fees. Seven Nordic TV channels could be broadcast in this way.[26]

The main argument in favour of NORDPROG is its economy. Investment costs are relatively low and since the operational costs are paid by its users they will be determined by market demand. However, NORDPROG is still only a proposal in front of the Nordic Council of Ministers. With the realisation that the original costs of the project had been underestimated, the Telecommunications Administrations have shown a greater interest in the Tele-X project and the possibility of combining it with national cable networks than in any further development of the NORDPROG proposals.

The discussions surrounding the advantages and disadvantages of Tele-X and the NORDPROG are very similar to the traditional debate about the need for DBS in countries with a high level of cable penetration. To understand the Nordic situation better it is therefore necessary to examine closely individual national cable policies and their links with international broadcasting by satellite.

SATELLITE BROADCASTING AND THE NORDIC AREA — POLICIES AND DEBATES

National policies with respect to cable television and satellite broadcasting in the Nordic countries have both similarities and differences. One very important similarity is the fact that the policies have developed as a response to external pressures; in most countries cable companies have forced the governments to take action, often before they have been fully prepared to do so. Another similar feature is that satellite policy has often included a desire to protect the national public service companies. A third point has been the interest in national industrial development. But there are many differences in the kinds of policies which have been developed. This section will examine the development of the Nordic policies with respect to broadcasting by satellite and it will focus on Sweden as a special case study.

Community antenna systems have been typical for Sweden since the mid-sixties. They were made compulsory when building new blocks of flats and were used mainly for receiving the two Swedish TV channels. Today about half of the Swedish population lives in flats with central antennas. In Denmark, central antenna systems were developed earlier, mostly by private initiative and often in small villages. The main reason was the possibility of getting additional TV channels from Sweden and/or Germany. Today, more than 50% of the Danes are linked to such central antennas. In Finland the penetration of central antennas is also about 50%, making it possible to watch Swedish and Estonian TV. In Norway, the percentage is somewhat lower. Here the cable networks were often established because of an interest in watching Swedish TV.[27] However, most of these systems have a limited capacity. They must be modernised if they are to be able to distribute a large number of satellite, or other, channels.

In 1982, the Swedish Telecommunications Administration presented a plan for the development of a national cable network. It proposed that the satellite broadcast services from abroad which had by then come to the attention of the Swedish general public should be distributed by cable. This proposal called for an immediate government commission to formulate Swedish policy in the general area of the new media. In fact, the Telecommunications Administration had long been active in the cable television field. In October 1983, it opened the first experimental cable network in southern Sweden. This network was used for the terrestrial reception of television channels from Denmark, East and West Germany and a number of radio channels.[28]

This service brought an immediate response and calls for the granting of licences for the reception of satellite broadcast services and their distribution through cable television networks. In the Spring of 1984, a provisional cable Act was passed by the Swedish parliament allowing the government to issue temporary licences for the reception of satellite broadcast services and their distribution through cable television networks. The licences were issued for an experimental period only and in anticipation of permanent legislation.

During this experimental period there was also growing interest in satellite broadcast services. In the Spring of 1984, 39 licences were issued. In May 1985, 82,000 households — about one percent of total households — had access to cable networks. Of these, 86 percent had access to satellite channels, mainly Sky Channel, Music Box, the Soviet Gorizont and the French TV5. In the border areas — which accounted for 60 percent of all cable networks — the cable networks also distributed Danish, Norwegian or Finnish TV received via terrestrial signals. About 80 percent of the cable

Figure 3.1: Transmissions in Swedish cable areas during the experiment period, share of cable households with access to different types of programming, autumn 1985 (per cent).

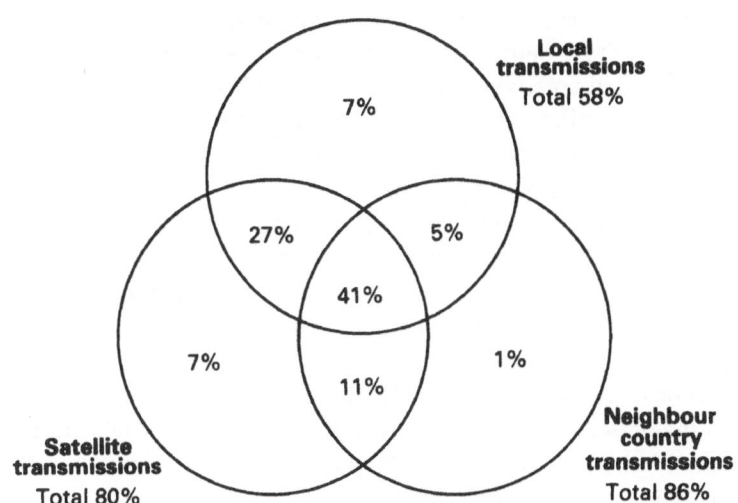

networks included at least one local channel, although the extent of local programming was usually very limited.

Most of the cable networks are owned and operated by the Swedish Telecommunications organisation; the remainder are either privately owned or are owned by cooperative housing companies. Generally, the growth of cable penetration during the experimental period was slower than had been expected. Instead of an anticipated 300,000 cabled households, the figure at the end of 1985 was only about 100,000, that is, about one per cent of Swedish households.

The government commission on satellite and cable policy appointed in 1982 published its final report at the end of 1984. The work had been characterised by a wish to reconcile the cultural and industrial aspects of cable and satellites. Within the commission there was a broad consensus concerning the importance of the industrial development: the expansion of the electronics industry and a stable labour market. The commission, however, did not think of satellite and cable as an important area of government planned policy as in, for example, France or Western Germany. It considered the establishment of cable as being mainly a matter of market demand. No technical specifications were recommended.

Thus the commission proposed liberal legislation: no restrictions should be imposed on the establishment of cable networks.[30]

The cultural aspects of satellite and cable generated much more political debate. The debate was dominated by two traditionally controversial issues (cf. section 2) — advertising and the freedom of speech. In these matters the commission reached a compromise. It recommended an advertising ban for local cable channels, while commercial satellite broadcasting was permitted, unless it carried advertising specifically addressed to a Swedish audience. Political opinion concerning the right to operate cable networks differed along a left-right dimension. The outcome was a compromise, recommending a formal licence to operate a network. Some restrictions were tied to the licence, among other things a ban on violence, racial discrimination and pornography in the programming. The control authority was given to a special government board — the Cable Board. In principle, the recommendations of the commission were transformed into a government bill, which parliament passed in 1985. The new cable legislation came into effect on January 1 1986.

The Swedish legislation pertaining to satellite and cable is often said to be one of the most liberal in Western Europe. It includes no rules for the establishment of cable networks. In addition, it is only valid for networks comprised of more than 50 (later extended to 100) households. Minor networks or individual households may receive any satellite channel. For the operation of larger networks a licence is needed. This distribution of satellite broadcasting may be handled directly by the network owner, but a licence to operate local cable channels can only be given to local cable companies. The licence is issued by the Cable Board. There is a 'must carry' rule which applies to the two TV channels of the Swedish public service company. Moreover, the ban on commercial programming, as well as the ban on violence and pornography in the programming, recommended by the commission, are included in the Cable Act.

With the new legislation enacted the Cable Board soon received a large number of applications for the new licences. As during the experimental period the Swedish Telecom was the most active cable company. By January 1987 about 200 licences had been granted, representing more than 200,000 households, covering about six percent of the Swedish population. Some areas, like the city of Gothenburg is highly cabled with a penetration of more than 20 percent of the households, but most parts of Sweden are not cabled at all. In the spring of 1987 the rate of increase in penetration has slowed down. It is also interesting to note that in this second wave of licence applications there were only a few which concerned local cable channels whereas almost all applicants wanted to distribute satellite broadcast services. The ban on advertising made local programming less attractive for the operators, e.g. local newspapers, which had been involved during the experimental period. They now saw few possibilities for financing a local channel.

As the largest cable operator Swedish Telecom, representing about 80 per cent of the cable households, set the standard for the packaging of satellite services. It offers a basic cable package, including Sky Channel, Super Channel (incorporating Music Box), the French TV5, the Soviet Gorizont, the compulsory national and local channels and in some areas the national channels of neighbouring countries (the recently ceased Europa TV was originally part of the package, CNN was included in the package in the spring of 1987). The monthly fee for this package is approximately $3 (20 SEK). Of this the Telecom charges about $1.50 for the programme distribution, the rest being, among other things, for technical installations and a Cable Board administrative fee. Households wishing additional satellite channels are offered Screen Sport, Lifestyle, Children's Channel and Arts Channel for a monthly cost of about $4 each plus a deposit fee for the decoder (about $65).

The cable legislation has meant that Sweden has been opened to satellite broadcasting from abroad. The legislation of the other Nordic countries reflects both similarities and differences regarding satellite broadcasting policy. At least on the surface, *Denmark* seems to have chosen quite another system. After the split in the NORDSAT discussions, Denmark set up a government commission to examine the whole media scene (see section 3). In 1983 the commission recommended the construction of a nationwide broadband network, the so called hybrid network. It is 'hybrid' because the network is meant to connect with the local central antenna systems.

Unlike Sweden, then, Denmark has opted for a planned policy for cable and in April 1984, its Parliament passed a bill to pursue these objectives so as to benefit from its industrial and employment implications. The financing of the network aroused some controversy, but a solution, based on a household fee, was agreed upon in late 1985. The decision in favour of the hybrid network implied a ban on the direct reception of satellite broadcasting from communication satellites, i.e. outside the hybrid network. The only exception to the rule applies to areas which have no opportunity to be connected to the network for the next two years. Licences for such areas are issued by the Ministry of Culture.[31]

By the end of 1986 there were still very few Danish households with access to satellite broadcasting. Interest in subscribing to cable networks has been much lower than expected. The annual subscription fee is quite large and this has aroused protests as well as the pirating of satellite broadcast services using illegal satellite reception dishes. The success of the satellite services has also been affected by the fact that a substantial majority of Danish households has access to television signals from the neighbouring countries through the community antenna systems, a fact which appears to make the immediate addition to new television channels less attractive.

The Danish Telecommunications organisation, which is responsible for the cable network, has tried to increase the penetration rate for cable and to reduce the activity of the pirates. The latter aroused a great deal of party political activity and led to a political decision and a compromise which now permits the use of satellite receiving dishes. It now appears that the Danish policy which started out as a very restrictive one will probably turn out to be as liberal as the Swedish one. The hybrid plan had originally gained much support because one of its main beneficiaries would have been the Danish industry. It later turned out, however, that most of the hardware would have had to be imported.

In *Norway* satellite broadcasting can be received by about a quarter of all households. The Norwegian law permits local cable companies to establish cable networks. The reception and distribution of satellite broadcasting is regarded as an exception to the sole right principle of the Norwegian national public service company. This exception is granted as a licence for the distribution of certain specified satellite channels. The licence is issued by the Ministry of Education and Culture and the total number of licences is limited. Being a country with only one channel, Norwegian households have always been interested in Swedish TV (the Norwegian and Swedish languages are very similar). As noted above, cable networks in Western Norway receive the two Swedish TV channels via satellite. Moreover, local cable channels as well as terrestrial local programming seem to have been very attractive in Norway. Another characteristic of the Norwegian system is the role

Figure 3.2: Summary of cable and satellite development in Sweden, Finland, Norway and Denmark

	SWEDEN	FINLAND	NORWAY	DENMARK
Cable development starts	1983	1974	1981	1983
Commissions on satellites and cable	1982	1979	1982	1980
Legislations on cable	1986	1987	1980 government may allow broadcasting in cable	1985
Concessions for cable issued by	Cable Board validity 3 years	Ministry of Communication validity 5 years	Department of Culture	A committee appointed by the Minister of Cultural Affairs
Cable ownership	Commercial basis	Commercial basis	Commercial basis	Commercial basis
Cable financing	By users	By users	By users	By users
Cable penetration 1986 (households)	206,000	225,000	450,000 (?)	17,000
(per cent)	6	12	26 (?)	1
Predicted penetration 1990 (per cent)	15–46	21–30	35	65
National broadcasting in cable	Must carry	Must carry	Must carry	Must carry
Distribution of commercial satellite channels in cable	Allowed if the advertising is not specifically addressed to the Swedish audience	No restrictions	No restrictions	As yet no satellite transmissions at all
Individual home reception of satellite channels	No restrictions	No restrictions	No restrictions	Banned
Population with access to terrestrial neighbour country television (per cent)	15	15	33	68
Households with video (per cent)	23	20	30	18

played by large private cable companies, the largest being Janco in Oslo.

Satellite broadcasting to Finland is regulated by a law enacted in 1987. Under it, licences for the distribution of satellite broadcast services are to be issued by the Ministry of Communication and are valid for five years. Cable systems are mostly built by local cable companies; these are most often owned by newspaper publishers and city councils, sometimes in cooperation with private telephone companies. The local cable operator must reserve a channel for local broadcasting and any surplus capacity on the network may also be used for this purpose. Rules regarding content are similar to those in Norway and Sweden although there is a greater emphasis on the need for local and regional productions. Such rules do not apply to the distribution of satellite broadcast services as long as these services are broadcast on separate and identifiable channels. However, the cable operator has to certify that the content of these channels are in general agreement with Finnish law. The reason for this exception to the cable Act is that when satellite broadcasting was first introduced there were no specific regulations in operation and the government commission proposing the new legislation did not want to reduce the freedom of the cable operators under the new law.

As can be seen in this brief overview of current regulations of communication satellite broadcasting into the Nordic countries, there are few restrictions — Denmark being a temporary exception — although there are differences in the methods of control. However, the actual reception of satellite broadcast services among the general public is still mainly dependent on the penetration of cable. The economic barrier is normally too high for a household to buy a satellite dish. So far there seems to have been limited opposition to the individual household fee for subscribing to a cable network. Swedish studies show that about 60 per cent of the households are willing to pay the normal fee (see above) and approximately a quarter are willing to pay twice the normal fee.[32] This willingness is somewhat lower in Denmark — about 40 percent consider the normal fee acceptable. Swedish households regard the Anglo-Saxon channels as more attractive; the percentage willing to pay for Sky Channel is greater than seven times those willing to pay for the French TV5. Perhaps more interesting is the fact that the willingness to pay for a local channel is not much higher than the willingness to pay for TV5.

It is hard to say what these figures imply about the growth of the cable networks and thereby the access to satellite broadcasting. As can be seen in Figure 2 the predictions for cable connections vary considerably both within (Sweden) and between countries. At the end of 1986 Norway had the highest penetration, but in 1990 Denmark is expected to be in the lead. But it has already been shown that the growth rate in the Danish system is lower than expected and early Swedish predictions have turned out to be much too high. It may be that in the beginning of the 1990s less than one third of the Nordic households will have access to satellite broadcasting through cable although there are a number of alternative predictions about cable's growth.[33]

THE AUDIENCE FOR SATELLITE BROADCASTING

The growth of cable networks in the Nordic countries has so far meant mainly an increase in the access to broadcasting from non-Nordic areas, in non-Nordic languages. In addition, national channels from neighbouring countries have been included in some areas. The changing TV situation has stimulated a lot of research, which makes it possible to draw some general conclusions. On the other hand, it is important to observe that different methods of data gathering have often been applied and this makes it hard to produce detailed comparisons between various studies. Furthermore, it has to be noted that the areas where satellite broadcasting was first introduced may not be typical of all Swedish households, e.g. being comprised more of blocks of flats in big cities than of small village areas.

As a general background it can be noted that the viewing of the national channels within the Nordic countries has been relatively stable during the last decade. With Denmark and Norway having one national channel each and Sweden having two, the TV audience on an average day amounts to about 75 percent of the viewing population, i.e. at least three persons out of four watch at least something on TV once a day. In Finland (two national channels) the figure is somewhat higher, 90 per cent. The average viewing time varies from 90 minutes in Denmark to approximately 120 minutes in Finland, with the Norwegian and Swedish figures lying in between.[34]

In areas where satellite broadcasting has been added to the existing national channels a general increase in viewing time can be

Figure 3.3: Shares of viewing time devoted to foriegn TV channels in some cable areas in Sweden 1984–5 (per cent).

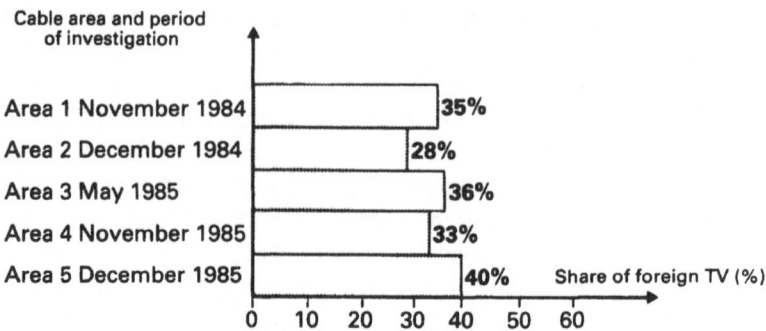

Notes: Area 1: Sky Channel, Music Box, TV 5 and Danish TV
Area 2: Sky Channel, and Gorizont
Area 3: Sky Channel, Music Box, TV 5 and Danish TV
Area 4: Sky Channel and TV 5
Area 5: Sky Channel, Music Box, TV 5, Gorizont,
          Childrens Channel, World Net, RAI 1,
          Mirror Vision and Danish TV

observed. In Sweden the increase ranges from 9 to 30 percent, depending on areas and number of available TV channels. The higher figure corresponds normally to cable areas with a larger number of satellite channels, but the rise in viewing time is far less than could be expected from the rise in broadcast time — normally the available programming has been many fold. Generally, it can be concluded that the more channels available, the less the use of the average channel. This conclusion is in accordance with earlier studies from the United States, where a fragmentation of the audience can be identified in multi-channel areas. However, it is important to stress that the potential for an increase in viewing time may be regarded as relatively high in the Nordic countries because the national channels usually start their transmissions in the late afternoon. It has to be added that the increase in viewing is often observed less than a year after the introduction of satellite programming. Thus, there might be an effect of the sheer novelty hidden in the figures.[35]

The role of satellite channels is illustrated in Figure 3.3. It shows the percentage of the TV viewing time devoted to satellite channels, including national channels from neighbouring countries in areas with satellite broadcasting available. Here the competition between the satellite channels and the Swedish national TV is evident. On the average, the channels from abroad account for more than one third of the viewing time. Corresponding results are found in Norway and Finland, although the percentages are somewhat lower — 20 percent. The reason, however, for the lower figures in Norway is that the Swedish national channels distributed through cable account for a substantial share of the viewing time. If we extend our comparison to channels from abroad, the Norwegian figures show that 'foreign viewing' accounts for about 40 percent of the viewing time. (In Denmark the share for TV from neighbouring countries — Sweden and/or Germany received with or without community antennas — is about 30 per cent.) A pan-European study of satellite TV viewing in areas where Anglo-Saxon channels were available showed that the Scandinavian countries accounted for the highest proportion of satellite viewing time, closely followed by the Netherlands. Switzerland, Belgium and West Germany ranked much lower for these channels. In the latter countries the viewing of TV from neighbouring countries played a more important role.[36] For Sweden the tendency seems to be the opposite: when satellite broadcasting was introduced the interest in national channels from the neighbouring countries decreased somewhat.

A general observation in the Nordic studies is that the viewers prefer the Anglo-Saxon satellite channels. The most popular one is Sky Channel with a weekly audience of about 75 percent of people able to receive it, i.e. at least three persons out of four watch at least something from this channel at least once a week. On an average day the percentage ranges between 20 and 33 percent. The higher figure is reported in areas where Sky Channel is one of two satellite channels and the lower from where it has a larger number of competing channels. These figures are taken from Swedish surveys but there are corresponding results from Finland and Norway. The second most popular channel is Music Box, whereas the French TV5 and the Soviet Gorizont range significantly lower. As has already been pointed out there might be a risk that satellite channels which have become available very recently may have the charm of novelty. One of the latest studies reported gives some support for such a hypothesis. After a couple of years Sky Channel seems to have lost some of its popularity, although it still is the most popular satellite channel.

Interest in the satellite channels varies a lot between different segments of the audience. Table 3.1 summarises data from a rather typical Swedish study conducted in 1985/86. It illustrates the total viewing time, the viewing time for channels from abroad, the share

Table 3.1: Average daily viewing time on Swedish and foreign television, share of viewing time devoted to foreign television and the weekly share for foreign channels in different segments of the audience. Cable area in Gothenburg, December 1985 (minutes/day, per cent)

Viewing time (minutes/day)	Gender		Age			Perceived English competence		VCR access		
	Male	Female	18–29	30–49	50–74	Bad	Good	No	Yes	Total
Swedish TV	105	102	93	92	124	121	92	105	102	103
Foreign TV	68	69	96	66	46	45	86	59	86	68
Total TV	173	171	189	158	170	166	178	164	188	171
Share of foreign TV (per cent)	39	40	51	42	27	27	48	36	46	40
Weekly audience share										
Sky Channel	76	74	92	76	60	59	87	68	89	75
Music Box	28	17	42	19	8	16	26	14	39	22
TV 5	5	5	5	2	8	7	4	4	8	5
Gorizont	15	14	9	12	22	20	14	14	14	14
Danish TV	16	44	41	42	69	59	49	49	55	51
n	125	158	88	104	91	115	163	186	97	283

of the total viewing time for these channels and percentages for weekly viewing for a selected number of independent satellite channels in different segments of the audience. Some general tendencies are easily observed. First, the interest in satellite broadcasting is significantly higher among younger people. A notable consequence of the introduction of the satellite channels is that younger groups, normally not very interested in TV viewing, have increased their viewing time more than other groups. Another important factor is the competence in English. People with a higher perceived competence in English watch significantly more satellite broadcasts. Also people possessing a VCR in their household tend to view more satellite TV. Although there is a slight correlation between age on one hand and competence in English and VCR access on the other, all three factors have an independent effect on satellite viewing.

The results must be interpreted both in terms of viewing preferences and the nature of the available channels. It is evident that the interest in satellite broadcasting in an early period of its existence shows the traditional pattern of all innovative processes: the more active you are, the more interested you are in trying new things.[38] Age and VCR access seem to be two indicators of the viewers' general interest in new media. A second and perhaps more striking conclusion is that extensive satellite viewing might be seen as a reflection of an Anglo-Saxon, or better Anglo-American, orientation. Both the role of the age factor and that of competence in the English language has little to do with the Anglo-American orientation, whereas this is important among the older generations. This is also the pattern of interest for Anglo-Saxon satellite channels: competence in English means nothing among the youth but a lot for people over 50 years of age. Thirdly, interest in Anglo-Saxon channels also means an interest in channels dominated by popular content and, to a very large extent, pop music.[39] It is not necessarily true for other types of channels, as can be seen in Table 3.1 for example in relation to the viewing of the national Danish television. The Anglo-Saxon channels present an alternative to the traditional programming of the public service companies.

In summary, it seems evident that a high interest in viewing Sky Channel or Music Box can be explained by an interest in an Anglo-American culture, especially an interest in Anglo-American pop music among the youth in the Nordic countries. If this conclusion is interpreted in terms of generational differences, we

Figure 3.4: Perceived functions of Swedish radio and TV, newspapers and foreign TV in two areas — one with cable, the other without — Sweden 1985 (per cent, index)

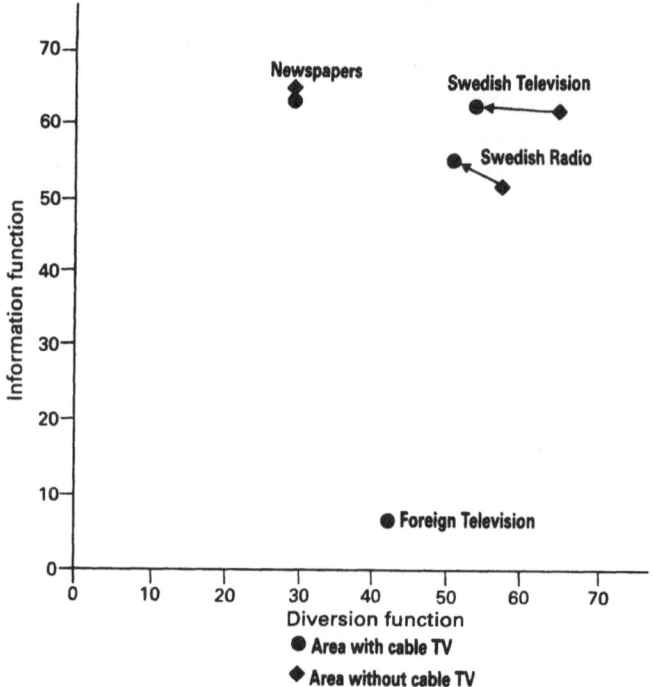

might expect a growing interest among the elderly over time. On the other hand, when such channels are no longer novelties, we might expect a decrease. The outcome, then, would indicate a viewing level of foreign channels similar to today's level, given the type of satellite channels available and little change in the media environment. Finally, it is important to stress that all figures presented here are taken from areas where satellite broadcasting is available. Since these areas comprise less than six percent of the Swedish population, satellite viewing, at least so far, is of limited importance to total TV viewing in Sweden.

The general media consumption in the Nordic countries has been very stable during recent decades. With the exception of a decline of the weekly press and cinema attendance, most media have had a firm share of the population. Studies of the consumption of other media than TV by people with access to satellite broadcasting have not shown any significant effects. However, a stability in usage does not necessarily preclude that an 'old medium' has been unaffected by a new one. There could, for example, be effects on the way of using it. To investigate this hypothesis, a detailed study on perceived media functions was conducted in a Swedish cable area in 1985/86.[40] The study was designed to match the cable area with a geographically similar area. In both areas the study concentrated on media use and attitudes to the media and there was also a comparison of the uses of different national media. The results showed that the evaluation of the media was spread across two main dimensions — the use of the media as sources of information and the use of the media for diversion.(Figure 3.4)

The black dots illustrate the perceived media functions in the area without cable and the rings give the results from the cable area. If we interpret the distance between the rings and the dots as the effects of the introduction of transnational satellite broadcasting like Sky Channel and Music Box (indicated as arrows) we can make some interesting observations. First, the satellite channels, not unexpectedly, have meant an addition in terms of diversion. Second, the national TV channels, and to a certain degree the radio, have lost a corresponding share and are now perceived as more informative. For newspapers no change in function is observed. A third observation, which cannot be seen in the figure is that the effects are much more evident in the younger generation, especially for radio. The conclusion seems to be that part of satellite broadcasting so far has shown a tendency to compete with radio, i.e. with secondary listening to pop music.

With satellite broadcasting from abroad as part of the future media environment, what is the potential role of the proposed Nordic channels from the Tele-X satellite? Is there also an audience for these Nordic TV channels? The question has been debated ever since the first NORDSAT commission. It is hard to evaluate the attractiveness of the coming channels. If we use results from studies of cable viewing we would expect the Nordic channels to play only a minor role in Sweden, with the exception of the immigrant groups from other Nordic countries. But we would predict a substantial viewing share for the Nordic channels in Denmark, Finland and Norway. The prediction reflects the traditional cultural interdependence among the Nordic countries, which is well illustrated in several studies of information flows. But there is another factor to consider: what will the national public service companies undertake in the new situation? This problem is tackled in the final section of this chapter.

THE YEARS TO COME: THE SITUATION FOR NORDIC
BROADCASTING

This chapter began with an analysis of the Nordic countries' approaches to the question of transnational broadcasting by satellite. It also explored the effects of these developments on the Nordic national broadcasting systems and on the audiences for national and international television services. This final section will provide a more detailed appraisal of the problems which the national public service broadcasting organisations face in the future and their strategies for overcoming those problems. It will also take note of the political environment in which these problems and their resolution will take place.

Four general tendencies have so far been observed. The first tendency in the Nordic approach to satellite broadcasting is that it started as a matter of Nordic cultural cooperation. Over time it also became a matter of industrial development on the national level. The conflict between the two perspectives reflects the traditional dualism of cultural and economic aspects in the area of media policy. In Section 2 this was illustrated by looking at the struggle between the 'cultural' and the 'commercial' alliances in the development of the Swedish media system during the last two decades. The strategies on the national levels differ somewhat among the Nordic countries, the market principle applied in Sweden and the conscious industrial effort initiated in Denmark being the two opposites.

Another tendency has been that political decision-makers have often found a national consensus on industrial policy but have diverged in their approaches to the cultural aspects of satellite broadcasting. Parties of the left have generally been more prepared to protect the public service broadcasting company as a bulwark against negative influences on the national culture. The cultural issues have often played a less important role, especially in contrast to the 'industrial imperative'. Advocates of the cultural aspect have been forced to present proof of their fears and not the other way round. The outcome has usually been an open policy for the new media, but with some 'cultural protection' built in, e.g. the 'must carry' principle, the ban on advertising or on violence.

On the whole then, the Nordic countries have adopted a fairly open policy to transnational broadcasting by satellite; a policy much influenced by a tradition of free trade between countries. As one critic observed 'if you cannot beat cultural imperialism, join it'.[41] However, although the legislation is also framed within an ideology of media social responsibility, it is in fact more liberal than the legislation which applies to the national public service organisations. The key issue remains the future of the latter organisations in the new media scene.

The last decade has seen a general turbulence within the national broadcasting systems of the Nordic countries. As has been argued, the first change was the development of local broadcasting, mainly in terms of local radio channels, but in Denmark also of local TV. Being on the defensive, the public service broadcasters have taken the challenge of continental satellite broadcasting seriously. Considering the very low penetration — so far less than three percent of the total Swedish audience on an average day — critics have claimed that there has been an overreaction. The public service companies, however, have argued that they have to make their long range plans and are awaiting a further growth in access to satellite broadcasting through cable development. Partly as a response to the new situation, partly as a result of the general turbulence in the media systems, there have been important changes in the national broadcasting systems. Denmark has decided to add a second TV channel. It is generally based on local stations and will be partly financed by advertising. In Finland there has been a decision to start a third TV channel. There has been a similar discussion in Norway, and a commission is going to examine possible organisational forms. In Sweden the national broadcasting company has proposed a third channel.

The most interesting aspect of these changes is the way in which they have come about, that is, either because of, or to meet the challenge of, broadcasting by satellite. The main motive behind a second Norwegian television channel was that it should offer 'an alternative to satellite broadcasting' and the Swedish government's proposal for national broadcasting, 1986-1992, was framed in terms of the need for public service broadcasting in a media scene characterised by transnational satellite broadcasting. Furthermore, the Swedish programme policy over the next few years calls for an earlier start to the daily programme schedules as well as for full programming at weekends: both key aspects of competition with satellite services. There are, in addition, some other common elements across the Nordic countries. Firstly, importance is attached to the national culture both for, and in, programming. The proposals in all the Nordic countries stress the importance of nationally produced television. Secondly, there is a tendency to call for an increase in the amount of television entertainment available so as to meet the challenge from abroad. These two strategies may seem contradictory but the aim is to produce high quality national entertainment.

It appears then that one of the most important effects of satellite broadcasting in the Nordic countries has been its influence on the national broadcasting systems. This influence, however, is not created by any massive audience reaction but merely by perceptions among political decision-makers and broadcasting planners. The same can be said of the ongoing debate about the possible introduction of advertising on Swedish TV. Representatives of what has been called 'the commercial alliance' have begun to use a new argument in favour of Swedish commercial television. The proponents of advertising argue that the existence of commercial satellite channels must result in an introduction of advertising in Swedish national TV; otherwise Swedish advertising money will go abroad. The argument has played an important role in the Swedish discussion on TV advertising, as it has in Denmark and Norway. This argument, although of doubtful validity, will probably be crucial for the expected decision on TV advertising in Sweden.[42]

It would be misleading to conclude that all the proposed changes in the Nordic TV systems have been determined by the introduction of transnational satellite broadcasting. A more fruitful interpretation seems to be that the new satellite channels have reinforced tendencies already existing in the Nordic countries. They have given legitimation for political views which earlier were considered with suspicion: a media system based on the market principle. Both in the national arena and transnationally the industrial interests have been advocating such a system.

Industrial groups in Sweden have proposed an independent TV organisation, financed by advertising and showing mainly entertainment programmes, as a response to the new satellite broadcasting channels. Such a channel is regarded as a commercially interesting alternative to the commercial channels available via satellite. These industrial groups have also been prepared to invest in such a television organisation. So far this has only come about in Finland where the electronic company Nokia owns 15 percent of the shares in the third television channel. The Swedish media industry has, in contrast, decided to invest abroad. The Kinnevik Investment company, through its subsidiary Medvik, has invested in the Luxembourg ASTRA satellite and is planning to transmit entertainment programming to Sweden this way. Medvik is also planning to broadcast a Nordic newschannel, Scansat, to the Scandinavian countries from 1988 onwards. Another Swedish media company, Esselte, has invested in Film Net, a satellite channel available on a pay-TV basis, as well as in a local Danish pay-TV station.

In the future, the role of the Telecommunications Administrations will also be important. To date, they have been active in the construction and operation of cable systems and they have signed agreements to market the transnational satellite channels. They have also taken up shares in some Pan-European satellite companies. They have yet to adopt a clear strategy between these different media: their interest in communications satellites and cable may make them less inclined to get involved in a DBS system, such as Tele-X. On the other hand, NOTELSAT, a subsidiary of the Norwegian and Swedish Telecommunications Administrations has been given responsibility for the operation of Tele-X. It would seem reasonable to suggest that they will attempt to combine the benefits of satellite and cable so as to develop and improve their telecommunications services. This is clearly important if they are also to meet the increasing competition in telecommunications from private services. One possible outcome of all this is that the telecommunications organisations will reduce their activities in the satellite broadcasting area.[43]

Our conclusion is that satellite broadcasting, or more important, the expectations tied to the development of satellite channels must be considered as a major factor behind the current turbulence in the Nordic broadcasting system. But we do not know the long term preferences of the audience. It seems probable that the interest in foreign broadcasting via satellite will persist at a certain level. Like video, it has a certain niche in the media system and will probably keep it. On the other hand, at the beginning of 1987, there was a slight tendency for audience opposition to cable to increase on account of the subscription charges. Nevertheless, despite this and the low level of cable penetration, the emergence of transnational satellite broadcasting has influenced the way of thinking about TV in the Nordic area: the satellites are here to stay and cannot be banned. Given this background it was very natural for the Swedish delegation at the Vienna meeting of December 1986 to conclude:

The need for cooperation and contact between the nations of Europe has increased as a result of the new internationalised media situation ..... attempting to strike a balance between protection of national culture and co-operation between European countries...[44]

On the other hand it might also be pointed out that, considering the new pressure on the public service broadcasting this has turned out to be very strong. For the foreseeable future it seems evident that the public broadcasters will dominate the Nordic media scene.

REFERENCES

1. Council of Europe. *European Ministerial Conference on Mass Media Policy, Vienna, December 9–10, 1986: sub-theme II:Public and Private Broadcasting in Europe. Report presented by the Swedish delegation.* Strasbourg 1986.
2. Schiller, H., *Communications and Cultural Domination.* International Arts & Sciences Press, New York, 1976.
3. Varis, T., The International Flow of Television Programs. *Journal of Communication* Vol. 34:1, 1984.
4. McQuail, D. & Siune, K. eds. *New Media Politics.*Comparative Perspectives in Western Europe. London—Beverly Hills—New Dehli, Sage, 1986, p. 197.
5. Examples of literature on Nordic policy on new media technology: Carlsson, U. ed. *UNESCO Consultation on Collaborative Research into the Impact of the New Communication Technologies.* Held at the University of Gothenburg January 27–30, 1986. Final report.
Hultén, O. *Mass Media and State Support in Sweden.* The Swedish Institute, Stockholm, 1984.
Kleinsteuber, H., McQuail, D. & Siune, K. eds. *Electronic Media and Politics in Western Europe.* Euromedia Research Group Handbook of National Systems. Frankfurt am M — New York, Campus,1986.
McQuail, D. & Siune, K. eds. *New Media Politics.* 1986.
Becker, J., Hedebro, G. & Paldán, L. eds. *Communication and Domination.* Essays to Honor Herbert I. Schiller. New Jersey, Ablex, 1986.
Nowak, Kjell, *Television in Sweden 1986.* Working Paper from the Centre for Mass Communication Research, University of Stockholm, 1987.
Valuable comments on the article have been given by Ulla Carlsson, Olof Hultén, Jan-Olof Gurinder and Karl Erik Rosengren.
6. Hultén O., Why NORDSAT — Why not? in *Media Culture and Society* 1981:3.
7. Ibid.
8. Wirén, K.H. *Kampen om TV.* Svensk TV-politik 1946–66. Stockholm, Gidlund, 1986.
9. Hadenius, S. & Weibull, L., *Massmedier.* En bok om press, radio, TV. Stockholm, Bonniers, 1985.
10. *Nordic Radio and Television via Satellite.* (1979) Final report NU A1979:4E. Hultén, O (1981). Why NORDSAT — Why not?, 1981.

11. *TV Across Borders.* NU 1974:19. *Nordisk radio och television via satellit.* (1977)Statssekreterargruppens slutbetänkande. NU A1977:7.

12. *Nordic Radio and Television via Satellite,* Final report, 1977, NU A1979:4E.

13. Hultén, O., Why NORDSAT — Why not?, 1981.

14. *Nordic Radio and Television via Satellite,* Final report, 1979. NU A1979:4E.

15. Svärd, S., *Med sladdar och satelliter.* Stockholm, Timbro, 1985.

16. Hultén, O., Why NORDSAT — Why not?, 1981.

17. Svärd, S., *Med sladdar och satelliter.* Stockholm, Timbro, 1985, p. 48. *Nordiskt radio — och TV — samt telesamarbete via satellit,* Rapporter från nordiska ministerrådets studiefas. Nordisk utredningsserie 1984:8.

18. Hultén, O., *Mass Media and State Support in Sweden.* The Swedish Institute, Stockholm, 1984, p. 47.

19. Dagens Nyheter, November 1981.

20. Hultén, O. Scandinavia: Nordic collaboration. In *Intermedia* 1986:4/5.

21. Ekecrantz, J. Policy Research and Research Policies in Communications, in Mosco, V. ed. *Policy Research in Telecommunications.* Proceedings from the Eleventh Annual Telecommunications Policy Research Conference, New Jersey, 1984.

22. Prehn, O., *Over kablerne neden satellitterne.*Telematikken — Kulturindustrin og den transnationale ekspansion. Ålborg, Aalborg universitetsforlag, 1985, p.20.
Kleinsteuber, H., McQuail, D. & Siune, K. eds. *Electronic Media and Politics in Western Europe,* 1986.

23. McQuail, D. & Siune, K. eds., *New Media Politics,*1986, p. 179.

24. Ekecrantz, J., 'State-Media Relations in Sweden', in Archer, C. & Maxwell, S. eds. *The Nordic Model.* Studies in Public Policy Innovations. Westmead-Farnborough, Gower, 1980.

25. von Feilitzen, C., *Nordiska satellitsändningar och publiken.* En forskningsöversikt och en prognos för Tele-X. PUB/Sveriges Radio, Stockholm, 1986.

26. Hultén, O., Scandinavia: Nordic collaboration. In *Intermedia* 1986:4/5.

27. Kleinsteuber, H., McQuail, D. & Siune, K. eds., *Electronic Media and Politics in Western Europe,* 1986.
von Feilitzen, C., *Nordiska satellitsändningar och publiken,* 1986.

28. Roe, K. *The Programme Output of Seven Cable-TV Channels.* A Descriptive Analysis. The Advent of Cable Systems in Sweden. University of Uppsala, Department of Sociology, Report No. 5, 1985.

29. Strid, I. & Weibull, L., *Mediesverige 1986.* Avd för masskommunikation, Göteborgs universitet, 1986.

30. *Via satellit och kabel.* Betänkande av massmediekommittén. Statens Offentliga Utredningar 1984:65,Stockholm, Liber.
Hultén, O., Current Development in the Electronic Media in Sweden, in Carlsson, U. ed. *Media in Transition.* Swedish Mass Communication Research on New Information Technology. A report from NORDICOM-SWEDEN, Göteborg, 1986.

31. Prehn, O., Community Television in Denmark — Three years of experiments. *The NORDICOM-Review* 1986:2.

32. Severinsson, R., *Publiken möter kabel-TV*. Avd för masskommunikation, Göteborgs universitet, 1985.
Djerf, M. *Funktioner hos kabel-TV*. Avd för masskommunikation, Göteborgs universitet, 1986.

33. *Effects of Television Advertising in Sweden*, Summary of Effects of Television Advertising (DsU 1986:2). The report of the Television Advertising Commission. The Swedish Ministry of Education and Cultural Affairs, DsU 1986:12.
Severinsson, R. (1987), *Den Nya medieframtiden — TV via satellit och Kabel*. Avdelningen för Masskommunikation, Göteborgs Universitet.

34. von Feilitzen, C., *Nordiska satellitsändningar och publiken*, 1986.

35. Severinsson, R., *Den nya medieframtiden*, 1986.

36. AGB, The Pan-European Television Audience Survey, 1986.

37. Hedman, L., Holmlöv, P.G. & Thomas, M., *Andra året med satellit-TV i Upplands Väsby*. Sociologiska institutionen, Uppsala universitet.
Roe, K., 1986. *Media and Social Life in Lund After the Advent of Cable Transmissions*. The Advent of Cable Systems in Sweden. University of Uppsala, Department of Sociology (in progress).
Severinsson, R. (1986). *Den nya medieframtiden*, 1986.
Sky Channel (1986) The Pan European Television Audience Survey, Vol. 1.

38. Rogers, E. & Shoemaker, F., *Communication of Innovations*. A Cross Cultural Approach, New York, Free Press,1971.

39. Roe, K., *The Programme Output of Seven Cable-TV Channels*. A Descriptive Analysis. The Advent of Cable Systems in Sweden. University of Uppsala, Department of Sociology, Report No. 5, 1985.

40. Djerf, M., *Funktioner hos kabel-TV*. Avd för masskommunikation, Göteborgs universitet, 1986.

41. Ekecrantz, J. (1980). 'State-Media Relations in Sweden', in Archer, C. & Maxwell, S. eds. *The Nordic Model*, 1980,p. 158.

42. *Effects of Television Advertising in Sweden*, 1986.

43. Gustafsson, K.E., *Radio och TV — Lokal och regional utveckling i Norden*, Svenska Kommunförbundet, Stockholm, 1986.

44. Council of Europe, *European Ministerial Conference on Mass Media Policy*, Strasbourg 1986.

4

Satellite Broadcasting Policy In West Germany—Political Conflict and Competition in a Decentralised System

Peter Humphreys

INTRODUCTION

In 1979 both France and West Germany withdrew suddenly from the European Space Agency's multipurpose telecommunications satellite project 'H SAT' (later known as L SAT) thereby marking the end of a common European satellite policy and a return to the pursuit of individual national ambitions. The Franco-German withdrawal also emphasised the extent to which their joint consortium EUROSATELLITE Ltd, heavily supported by the state in each country, was likely to make the running in DBS initiatives in Western Europe.

The growing importance attached to satellite broadcasting policy by both countries, but by West Germany in particular, will be the focus of this chapter. Centrally concerned with DBS, it also examines how established patterns of broadcasting have already been convulsed by 'point-to-point' satellite broadcasting, namely communications satellite linking with the country's growing cable infrastructure. In particular, it focuses upon the highly political nature of the issue of satellite broadcasting in West Germany.

FRANCO-GERMAN TECHNICAL AND INDUSTRIAL COLLABORATION IN SATELLITE CONSTRUCTION

Franco-German collaboration in space-satellite construction has a fairly long history. From the mid-1960s on until the mid 1970s, both countries devoted much aerospace energy to the joint construction of a series of communications satellites, called SYMPHONIE. The SYMPHONIE satellites (launched in August and December 1974) were built by a consortium of respective 'national champions': Messerschmidt Boelkow Blohm, Siemens, AEG, SAT, SNIAS ('Aerospatiale') and Thomson-CSF. On the West German side alone, the project had absorbed DM 600 million of state support from the federal government.

During this extended period, both countries developed an advanced capability in satellite technology—a capability that led them to become increasingly reluctant to pool their expertise in pan-European projects. The 1979 decision to withdraw from 'H-SAT' did not, therefore, come as a surprise. Until then, growing West German interest in the broadcasting applications of the latest satellite technology had led the Federal Republic to contribute the largest individual national share amounting to 32% of the 'H-SAT' budget (closely followed by France's share of 28%). The completion of the SYMPHONIE series in the mid 1970s, however, had released such rich satellite construction capacity in both countries that they were able to seriously entertain going it alone instead of in collaborative DBS development, then coming to be widely seen as the next step in commercial satellite developments.

The formal explanation given for quitting the ESA project related to the 1977 WARC agreement, which by allocating geostationary orbit positions and frequencies on a national basis to individual states was held to have removed the basis for a Europe-wide common project such as 'H-SAT'.[1] However, the major reason was undoubtedly the growing ambition of both countries to exploit their technological lead (within Europe) in the race to commercialise broadcasting satellite technology. In particular, they wanted to capitalise upon the expertise and capacity gained in the recently completed collaborative SYMPHONIE project. The argument that the WARC regulations made it incumbent upon individual nations to develop their own satellites was clearly specious; yet it served as a sufficient pretext for pressing on with particularist ventures in the race to produce demonstration models for a future export market that was held to be highly promising.[2]

In West Germany, the initiative for DBS development can be seen as the result of a confluence of state and industry aims, reflecting the general tendency during the 1970s of the state to intervene, far more than had been hitherto customary in the post-war West German economy, in order to anticipate the direction of industrial change and achieve economic restructuration. Satellite development was a field of activity (within the area of communications technologies more widely) that had been seriously considered to be interesting for its future export potential in the spirit of, and according to the guidelines laid down in, the West German Social Democrats' (SPD) 'structural modernisation programme'.[3]

Following the WARC 77 allocation of DBS orbit positions, specific satellite technology studies were hastily conducted by the West German Institute for Aerospace (DFVLR) for the Federal Ministry of Research and Technology (BMFT). This Ministry had actually been established by the SPD in 1972 in explicit recognition of the importance of state support in the processes of research, development and introduction of the new technologies. The BMFT was for a while, under the SPD, the central force for its 'anticipatory' technology policy. The logic of engineering collaborative adjustments to new international competitive pressures—like those unleashed by DBS technology — was also particularly characteristic of Chancellor Helmut Schmidt's conception of 'Model Germany', a technocratic version of Social Democracy that gave priority to the international economy.

Continued collaboration with the French meant that the risks and costs of development could be shared, while at the same time progress would be speedier—an important calculation in the race to be first to launch a demonstration model. Strong support for DBS development came from the BMFT and the DFVLR. Rather less enthusiasm initially, but a certain interest in keeping abreast of any development that concerned its telecommunications monopoly, came from the powerful, but 'conservative' Federal Ministry for Post and Telecommunications (BMPT). The possibility of continued space technology collaboration with the French was naturally warmly welcomed by a number of leading West German manufacturing firms with interests in aerospace, electronics and telecommunications—notably, Messerschmidt Boelkow Blohm (MBB), AEG-Telefunken, Dornier and SEL.[4]

French industry—notably, Thomson-CSF and SNIAS ('Aerospatiale')—also welcomed this collaboration. It was applauded within the CNES, the French space research body, and commended by the French Ministry of Industry. At highest political levels, the French were even keener than the Germans—who appeared initially to be interested only in the immediate future—to make the project the basis for continued long-term collaboration in space technology. The political will to cement the 'Paris-Bonn axis' was probably stronger on the French side, while the economic rationale for sharing expertise and the immediate costs of DBS development was shared by both sides in equal measure.

As a result of urgent discussions during 1979 between the West German BMFT and the French Industry Ministry, the French and West German governments signed a formal treaty on technical and industrial cooperation in the field of DBS in April 1980.[5] The first step was to be the construction of two identical DBS satellites—TDF I (French) and TV SAT (West German)—which would be placed in geostationary orbit by the French (ESA) ARIANE rocket and thereafter managed by the respective national telecommunications authorities, Télédiffusion de France (TDF) and the Bundespost. Later, back-up satellites might be constructed to render the new DBS systems fully (rather than pre-) operational. Decisions on these back-up satellites were to be made independently by the respective national authorities. In the event, the French government was to decide positively after many delays and uncertainties; in December 1984 Socialist Prime Minister Fabius announced the French intention to build a second satellite, TDF 2, and to launch it by the end of 1988. However, such decisiveness was more difficult to achieve on the West German side; by late 1986 the West German Bundespost Minister, Christian Schwarz-Schilling, had still not actually put his signature to the contract for a second satellite—for reasons that will become clear from the rest of this chapter.

The project was to be administered by a management committee made up of executives and specialists from the relevant Franco-German bodies (TDF, CNES, BMFT, BMPF) and, in July 1982, the Franco-German consortium EUROSATELLITE Ltd was formally contracted to construct the satellites. EUROSATELLITE Ltd, with its headquarters in Munich incorporated MBB (24%), AEG-Telefunken (24%), SNIAS ('Aerospatiale') (24%), Thomson-CSF (24%), and a Belgian firm ETCA (4%). Delivery date for the satellites was initially set for mid-1985. The estimated cost of the satellites was initially set at DM 225 million for TV SAT, and FFr 555 million for TDF 1—with an additional cost of DM 115 million for each launch by ARIANE. This huge investment—the lion's share borne by the state in each country—was to be offset by capturing a large share of a world market the volume of which was estimated variously to be between DM 25-50,000 million.[6] The first 'customer' arrived almost immediately when the Swedish government decided that EUROSATELLITE Ltd should build the Swedish (later

Scandinavian) Tele-X DBS satellite, with a 'token' participation by Swedish manufacturers.

THE 'TECHNOCRATIC' MODE OF LAUNCHING THE PROJECT AND ITS CRITICS

At this stage, it should be noted that these DBS initiatives started life purely as a result of decisions relating to technology and industrial policy, on the one hand, and of perceptions of a developing international market for communications hardware, on the other. Little consideration seems to have been given to the consequences for broadcasting policy. On this count, serious reservations had been expressed by SPD media-policy makers in the Federal Chancellor's Office, notably by the fierce opponent of commercial television Albrecht Müller.[7]

Müller saw that the introduction of 'new media'—and the corresponding multiplication of broadcasting frequencies—strengthened considerably the cause of the increasingly vocal commercial television lobby in West Germany. Furthermore, according to Müller, the new media threatened to become the lever for the eventual effective total privatisation and commercialisation of the entire West German broadcasting system. As will be seen, the established post-war West German system was a deeply entrenched 'public service monopoly' and Müller represented a fairly generalised fear within the SPD that once introduced commercial television would steadily undermine the quality of public-service television.

At this time, official SPD media policy reflected predominantly this line. It was firmly set against introducing any commercial television whatsoever. Helmut Schmidt himself was greatly influenced by Müller's arguments against cable television though he did not seem to appreciate DBS in the same light. The most likely explanation for this was that with only three new channels (in the pre-operational stage, only later extended to five), TV SAT was generally expected to be allocated, at least with the SPD in power, to the existing public-service broadcasters, ARD and ZDF. Officially, the SPD-led government merely committed itself 'not to prejudice the development of media policy' by the project, whose main value was assumed to be its 'demonstration' effect on future overseas customers. It was also widely regarded as being only of an 'experimental' nature.[8] There was also some talk of using it for cultural-exchange purposes between France and West Germany. By and large, though, at the time of the collapse of the SPD/FDP 'social-liberal coalition' in Bonn, in October 1982, there were still no clear indications about the use to which TV SAT would actually be put.

THE COMPLEX QUESTION OF BROADCASTING REGULATION

The issue of satellite broadcasting regulation proved to be a particularly complex matter—impeding developments—in West Germany for two main, inter-acting reasons. Firstly, the federal system divides telecommunications policy (a federal matter) from broadcasting policy (a field largely reserved to the individual states or *Länder*). Secondly, media policy, particularly policy for the 'new media', became, during the 1970s, increasingly the object of an intense political polarisation between the major political parties.

The *Länder* of the Federal Republic are constitutionally assigned responsibility for cultural policy and, significantly, this includes competence for broadcasting policy. As a result, there is a significant sub-national arena for partisan conflict between the parties over broadcasting and different policies may be advocated and pursued by individual *Land* governments. Moreover, there is a significant hurdle for centralised policy making in a matter that combines industrial/technological policy—DBS being just such a centrally launched policy originally—with cultural policies for broadcasting.

It might be logically expected to follow from this simple structural division of interests and competences that the federal authority—in particular, the Bundespost—will be content to promote industrial and technology policies for telecommunications; while the *Länder* will be concerned with broadcasting aspects of policy. After all, there would be a kind of 'functional simplicity' to such a state of affairs.

This was indeed the pattern for most of the post-war period but developments concerning the new media were much more complex. Where there is a political will to steer national broadcasting developments—as became the case with the new media— the Bundespost presents itself as a powerful agent able to exert a national influence—particularly through its authority over, and ability to direct, telecommunications investment (e.g. cable, satellite reception equipment, relay stations, frequencies etc) within and between the individual *Länder*. Similarly, the *Länder* governments' interests in broadcasting policy are not simply confined to the cultural aspects of policy, but also cover broader economic considerations—such as the question of attracting inward investment in infrastructure and media-production (hardware and software) — a matter of heightened concern during a period of recession and industrial change.

In fact, due mainly to the convergence of telecommunications and broadcasting as a result of the 'new media', policy makers at *Land* level have found themselves increasingly compelled to act as brokers between these various aspects of policy. At the same time, the 'new media' have been the catalyst for a break-down of an established inter-*Land* consensus about broadcasting. In addition to ideological divergence, this has unleashed a destabilising dynamic of inter-*Land* competition against the background of the erosion of the hitherto dominant notion of *Land* sovereignty in the field of broadcasting policy. In turn, this has had adverse effects on the development of the new broadcasting sector.

The at-first-sight baffling complexity of the West German case might well be easier understood by outside observers if considered as an interesting microcosm of wider European developments, where the traditional sovereignties being broken down by satellite broadcasting are those of national not sub-national units. Nevertheless, disentangling the confusion requires an understanding of the general outlines of the post-war development of broadcasting in West Germany.[9]

THE POST-WAR CONSENSUS

During the post-war period, the West German *Länder* evolved a system of 'cooperative federalism' in broadcasting affairs. At the root of this 'cooperative federalism' was a common interest in, and commitment to, the principle of a public-trustee structure of public-service broadcasting. This commitment had several mutually reinforcing features.

There was a common concern to secure broadcasting against capture either by the state or by sectional interests in society. Strict public-service regulation and the structures of the broadcasting organisations themselves reflected a powerful normative constitutional imperative 'to regulate for pluralism' and thereby assure 'balance' and 'diversity' in broadcasting. In view of both technical considerations (scarce frequencies) and the very high financial costs of television production, it was assumed that pluralism, balance and diversity could not be reflected adequately between, but only within, a necessarily relatively restricted number of broadcasting organisations. Moreover, these broadcasting organisations had to be democratically accountable.

The founders of the Federal Republic's broadcasting system decided against a private law form of broadcasting organisation, instead they chose in favour of making them non-profit-making, self-administered 'corporations governed under public law' (Anstalten des öffentlichen Rechts). This legal form guaranteed the legal autonomy, independence of programme planning and production, and financial self-sufficiency of the broadcasters. Financial self-sufficiency and self-administration were to be the insurance against capture by sectional interests. At the same time, governments were not to direct broadcasters, nor intervene in the broadcasting of programmes, except in the case of contravention of broadcasting laws or other laws.

Within each public-service broadcasting organisation an 'Administrative Board' was responsible for business management, while an *Intendant* ('Director General') fulfilled the key executive role and represented the corporation in law. However, most importantly, both the Verwaltungsrat and the *Intendant* were accountable to a third internal organ of control, the *Rundfunkrat*. The *Rundfunkrat* was a supervisory organ that gave public representation to the field of broadcasting. In some broadcasting corporations, its members were chosen by the elected parliaments of the *Länder;* in others, they were elected or delegated by a whole range of 'socially relevant groups' such as churches, trade unions, universities, political parties etc. In most broadcasting organisations, the *Intendant* and all or most members of the Board were appointed by the *Rundfunkrat*, which also supervised programming.

This kind of regulation was called 'internal control' *(Binnenkontrolle);* control was thus internally built into the very structure of the independent broadcasting organisations themselves. The main function of the *Rundfunkrat* was to assure 'internal pluralism' *(Binnenpluralismus),* that is, a balance of opinion and a fair representation of social diversity within each channel. During the post-war period there arose a basic consensus on this form of regulation, backed up by interventions of the Federal Constitutional Court, whose judgements were interpreted as hostile to commercial broadcasting. The broadcasting consensus thus seemed secure.[10]

In spite of the well-defined principles and mechanisms for ensuring pluralism within broadcasting, the public-service corporations have been subject to considerable party influence. In practice, the broadcasting landscape in West Germany came to reflect the country's political geography—with an SPD bias largely confined to the northern half, and a Christian Democratic/Christian Social bias (CDU/CSU) mainly in the southern half—modified by some respect for the principle of Proporz ('proportionality', providing for a notionally shared influence). For most of the post-war period the central objective of the political parties' media policies has been either to gain, or to extend, their control of the broadcasting corporations. Nevertheless, for many years a 'hidden consensus' existed between the parties on the desirability of (and rough equitability produced by) this kind of 'gamesmanship', reinforcing the consensus on broadcasting. However, the CDU/CSU was always resentful of the alleged ARD bias towards the SPD and in the 1970s it became increasingly discontented with the existing state of affairs.

The practice of cooperative federalism was strengthened further by the shared suspicion of the *Länder* that the federal government harboured an ambition to encroach upon their broadcasting sovereignty. Significantly, in the period 1957-60 in the face of attempts to introduce a federal commercial channel in Bonn, both Christian Democrat/Christian Social and Social Democrat governed *Länder* alike rallied in defence of their *Länder* prerogatives.

The consensus among the *Länder* was further cemented when the Federal Constitutional Court ruled in their favour. Broadcasting, as an element of *Kultur*, was declared categorically to be within the jurisdiction of the *Länder*. Significantly, the Court drew a fundamental difference between the nature of the press and broadcasting, seeming to reject the possibility of private, commercial broadcasting on the grounds that in 'the field of broadcasting the number of those providing services must remain necessarily relatively small, both for technical reasons and in view of the exceptionally high financial expenditure involved'. Moreover, in 1971 the Court seemed once again to confirm the public-service monopoly of broadcasting. Article One of an important 1971 Ruling specified that '...the activity of the broadcasting organisations takes place in the public domain. The broadcasting organisations are publically accountable; they perform tasks of public administration; they fulfil an integrating function for the whole of the state. Their activity is not of a commercial kind'.[11]

Lastly, there was powerful economic support for the consensus: television production was hugely expensive. The broadcasting organisations of many 'city-states', e.g. Hamburg, and those of small, less affluent *Länder*, depended upon mutual inter-*Land* cooperation in one form or another. For most of the post-war period, the combined constraints of the existing technology, the high financial cost of broadcasting and the very strict regulatory climate meant that the conditions were largely absent for a powerful and successful challenge to be mounted from the private, commercial sector.

STRUCTURE OF THE ESTABLISHED PUBLIC-SECTOR BROADCASTING SYSTEM: THE KEY ROLE OF THE 'STAATSVERTRAG'

The precise organisational and legal structure of the broadcasting corporations was defined by *Land* laws and, if more than one *Land* were concerned, by means of an inter-state treaty (*Staatsvertrag*) between all or several *Länder*. Altogether nine regional public-service stations were set up.

In 1950, these organisations, plus two federal radio broadcasting corporations, all combined in the Arbeitsgemeinschaft der Rundfunkanstalten Deutschlands (ARD) to produce the first national television network. In addition, the broadcasting corporations produced so-called 'third' television channels, transmitted regionally. The ARD represented the common interests of the corporations, settled joint programming concerns and coordinated joint technical, legal and commercial arrangements. In addition, it concerned itself with tackling new developments in broadcasting. One such policy initiative was to be the establishment of a new satellite broadcasting channel ('Eins Plus'). The country's second original, national public-service channel was produced by a public corporation established in 1961, largely as a result of the abortive attempt to establish a federal, commercial television service. Zweites Deutsches Fernsehen (ZDF) was a more 'centralised' structure than the ARD. It was based in Mainz and was commissioned to produce an 'alternative' to the first national channel. ZDF was also to meet the challenge of satellite broadcasting by producing its own, public-service satellite service ('3 SAT'). Despite its independent, 'national' character, ZDF was the product of a specific *Staatsvertrag* between the Prime Ministers of the individual *Länder*.[12]

Cooperative federalism has also been a precondition of the mode of finance of the public-service broadcasters. Advertising accounts for a very strictly limited amount of the public-service broadcasters' income. The greater part of the public-service broadcasters' finance (approx 70%) comes from licence fees. More importantly, the level of the licence fee (currently DM 16.25 per month) is fixed by the council of the Prime Ministers of the *Länder*. Thus the very financial underpinning of the West German broadcasting system depends on accommodation and agreement between the *Länder*.

THE POLITICAL STRUGGLE: THE DEPARTURE FROM CONSENSUS

Thus during the 1950s and 1960s, the politics of broadcasting were characterised by the patronage of the 'party state', but at the same time bounded by an emerging consensus about the decentralised and public-service nature of broadcasting regulation. The issues raised by cable and satellite broadcasting have caused a wide departure from this post-war consensus, particularly with regards to the public-service nature of broadcasting regulation. During the 1970s, the West German conservative parties—the CDU and the CSU— and the liberal party, the FDP, began to herald cable and satellite television, increasingly enthusiastically, as an opportunity to introduce commercial broadcasting in the Federal Republic. They embellished their arguments with sweeping claims about 'freedom' and 'democracy'. One new slogan claimed that the new media would bring greater 'freedom' for the 'politically mature and responsible citizen'. Another powerful slogan carried the message that 'more programmes means more choice, not more watching'. The message was: 'more programmes, more choice; more competition, more freedom'. Indeed, the CDU/CSU media policy statements even made explicit the connection between this equation and an improved democracy.[13]

However, concealed barely behind this noble concern for the viewer's 'freedom' lay an unashamed political motivation. The

117

CDU/CSU's support for commercial broadcasting reflected, at least in part, the belief that its values and political tone would be more in tune with their own political positions, as was clearly the case with much of the private, commercially organised press. Therefore, during the 1970s the CDU/CSU mounted an increasingly vigorous campaign in favour of commercial broadcasting and the new media. Such developments also offered an opportunity to promote the economic and industrial futures of their own regions. The CDU/CSU received massive support from West Germany's generally conservative, powerful newspaper and magazine publishers, who anticipated combining a massive extension of their editorial 'reach' with a new area of lucrative enterprise.[14]

In contrast, the SDP, the dominant party of government in Bonn from 1969 until 1982, was extremely concerned about the concentration of media power—and in particular, the concentration of media power in the hand of right-wing publishers like the Axel Springer concern. In the early 1970s it had developed a fairly radical media policy which sought to 'democratise' the media but by the end of the 1970s the SPD policy had lost all traces of its former 'radicalism'. Instead it had become very defensive in face of the mounting CDU/CSU campaign and the obvious interest of the press barons to diversify into the electronic media.

While concerned to develop the new technologies, the SPD totally rejected deregulation and commercialisation of broadcasting. By the early 1980s the SDP was pledged to staunchly maintain the existing regulatory principles and mechanisms and to resist any commercialisation of broadcasting and media-concentration. This stand by the SPD was supported by a broad coalition of forces: the existing public-service broadcasters, the churches and the trade unions.[15]

THE CONSTITUTIONAL COURT SETS NEW PARAMETERS TO THE DEBATE

Both sides contrived to draw encouragement when in 1981 the Federal Constitutional Court suddenly formulated a number of principles that clearly anticipated dramatic structural changes in broadcasting and sought to guide future broadcasting regulation in the age of cable and satellite. To the delight of the SPD, the Court reiterated the normative constitutional imperative to regulate for pluralism, balance and diversity, specifying that, even if cable and satellite technologies removed the technical rationale for a public-service monopoly (scarce frequencies), broadcasting should not be left to 'the free play of market forces'.[16]

However, the partisans of commercial television welcomed the fact that the Court now suggested a possible alternative model to 'internal pluralism' for regulatory policy. According to a new, so-called 'external pluralism' model, balance and diversity might be achieved by a greatly increased number of channels and programmes: these could be permitted to be individually biased so long as, in totality, they reflected a balance. The notion of 'external pluralism' amounted to a recognition by the Federal Constitutional Court of the possibilities for the development of a new, 'market' model of broadcasting regulation. It recognised that, in view of the new (feasible) multiplication of channels, it was now possible to expect that the principles of pluralism, balance and diversity might be better achieved by permitting more access to the medium. In fact, it was an admission by the highest authority in the land that broadcasting could develop into a kind of 'electronic publishing' so long as the principles of pluralism, balance and diversity were assured.

For an initial period at least, the SPD continued to argue fiercely for 'internal pluralism' and 'internal control'. The SPD feared that 'external pluralism' would lead to the operation, in practice, of a kind of cultural Gresham's law: low quality television with a bland (naturally conservative) 'neutrality' would drive out high quality television that dared to tackle controversial issues in an informed and diverse manner. Moreover, cheap entertainment programmes would predominate as private and public broadcasters alike were compelled to go 'down market' in ever fiercer competition for viewer ratings. The result would not be diversity; rather it would be 'more of the same' and 'lowest common denominator' television. On the other hand, the CDU/CSU maintained that the new model would ensure pluralism, indeed an improved pluralism by dint of the extended consumer choice: the assumption was that the greatly expanded 'menu' of television channels would bring about a 'quasi-automatic' pluralism, balance and diversity.

THE CHANGE OF POWER IN BONN

While the Bundespost remained under its control, the SPD was able to turn federal competence for telecommunications to block widespread cabling—only four new, 'experimental' or 'pilot' projects were envisaged to complement the country's existing patchwork of 'low-tech' cable islands (developed to cover areas of poor 'off-air' reception). At the same time, and in the same way, the SPD was able to restrict the use of communications satellites purely to data and telephone links. However, the direction of West German media policy changed abruptly after the collapse of the longstanding SDP/FDP government and the subsequent confirmation in federal power, by the general elections of March 1983, of a new 'conservative/liberal' CDU/CSU/FDP government. Although broadcasting policy was traditionally (and constitutionally) an area of *Länder* competence, the Bonn 'change of power' had an altogether new impact on broadcasting policy. The new federal Minister for Posts and Telecommunications, Christian Schwarz-Schilling, had long been a leading campaigner for commercial broadcasting. The occupation of this particular ministerial office in Bonn by a CDU media policy specialist suggested that telecommunications policy might now be enlisted in the cause of CDU/CSU media policy.

Almost immediately, Schwarz-Schilling launched an ambitious and costly programme to cable the country. He proclaimed openly that its prime aim was to 'revolutionise broadcasting'. Moreover, he emphasised that cable investment by the Bundespost would henceforth be channeled to those *Länder* which enacted new media laws which would permit commercial exploitation of this 'revolution'. Subsequently, Schwarz-Schilling was to threaten repeatedly to suspend cabling in SPD-governed *Länder* that resisted introducing new legislation to allow private commercial television. Also, at the end of 1983, he announced that the Bundespost would release extra capacity for satellite broadcasting: in addition to the four DBS television channels (plus one devoted to radio) on TV SAT, two channels on the ECS-1 communications satellite, six channels rented on an Intelsat satellite; and these latter would later be replaced or complemented by seven channels on a new West German communications satellite system called DFS KOPERNIKUS. All this amounted to a colossal investment by the Bundespost. The cable programme alone was allocated DM 1,000 million per annum (rising to 1,500m in 1986). In addition, Eutelsat charged around ECU 1.8 million per year for the use of ECS-1 capacity (the Intelsat channels came somewhat cheaper). The plethora of additional channels would, it was hoped, stimulate the speediest possible development of a West German commercial broadcasting industry.[17]

THE NEED FOR A NEW *'STAATSVERTRAG'*

Following the new ruling of the Constitutional Court and the change of power in Bonn, the CDU/CSU *Länder* quickly prepared legislation enabling commercial television. Overall, the general drift was inexorably away from the 'internal pluralism' model of regulation towards 'external pluralism', as CDU/CSU policy makers at the *Land* level sought to create a suitably deregulated environment for new entrepreneurial, multimedia activity in their *Länder*.

However, by its very nature satellite television, especially DBS, would amount to 'national' broadcasting in that it could be received by the whole country. Therefore, the heads of government of the *Länder*, including those of the SPD *Länder* had now to decide collectively about the allocation of at least ten potential satellite television channels. As importantly, the heads of government had also to agree common regulatory principles for the 'feeding' of cable systems with programmes that originated in other *Länder* where different regulation would be operative. New common principles for advertising on the satellite channels would also be required — since advertising revenue would be the 'life-blood' of the new broadcasters and therefore its regulation had repercussions beyond any particular purely regional jurisdiction. In order to arrive at an arrangement for such matters that commonly affected all *Länder* traditional practice suggested that negotiation of a new *Staatsvertrag* would be necessary. This process required a minimum of consensus; the alternative would be the possible disintegration of the Federal Republic's broadcasting landscape. As one West German broadcaster put it, the country would amount to a '...mosaic of individual media territories, a truly paradoxical situation in the day and age of satellites'.[18]

THE LONG AND DIFFICULT PATH BACK TOWARD POLITICAL COMPROMISE

SPD policy makers in the *Länder* were now faced with very difficult choices. It was clear that CDU/CSU policy was explicitly

121

designed to attract new broadcasting entrepreneurs to establish their centres of activity and invest in CDU/CSU *Länder* and the SPD policy makers were fearful lest their 'fundamentalist' opposition to commercial television would push willing investors into alliances with foreign broadcasters and into using foreign satellites. This would entail a loss of investment, without in any way protecting either West German culture or West German advertising markets. Indeed, these markets were being tapped already by foreign channels such as the British-based SKY Channel, which was 'feeding' West German cable systems from a transponder on ECS 1.

These fears were also confirmed when in 1981 the West German newspaper publishers' federation showed an interest in taking a share in the LUX-SAT project and in late 1982 when the huge West German multimedia conglomerate Bertelsmann announced that it was to take a share of 40% with Compagnie Luxembourgeoise de Télédiffusion (CLT) in a new commercial German-language channel, RTL-Plus, to be broadcast into West Germany from Luxembourg, where it would be immune to West German regulatory policies.

In due course, there were calls for a new realism in SPD media policy and, at its party conference in 1984,the national SPD adopted, by a narrow majority, a new media action programme. The new programme effectively conceded that fundamental opposition to commercial television would be ineffective and self-defeating.[19]

Threats from CDU *Ministerpräsidenten* 'to go it alone' and so undermine the unity of the West German Broadcasting scene if an agreement for cable and satellite broadcasting could not be reached also played a part in shifting the SPD's position over the new media. In each *Land*, however, different forces were brought to bear on the negotiations for new agreements and laws on the new media.

Nevertheless, by the end of 1985, deregulatory new media laws had been enacted in the CDU/CSU states and also in SPD Hamburg.[20] The new legislation in SPD states reflected the SPD's changed media policy and amounted to a considerable measure of deregulation. Hamburg's new media law, for example, seemed to accept the new regulatory formula of 'external pluralism' and 'external control'. However, it attempted to combine this with 'internal pluralistic' elements. The law specified (article 19) that 'any' applicant for a regional franchise, who already had a dominant position in the city's daily newspaper market, should not himself personally be eligible for a private broadcasting licence; moreover, his share in any company receiving a licence should not exceed 25%. This level was clearly fixed above the direct shareholding of the Springer Press in the new private, commercial satellite channel, SAT 1, that had meanwhile been franchised in the Rhineland-Palatinate (more of which later).

The draft-law produced in North-Rhine Westphalia in February 1985 preparing the way for private commercial television—and based on the changed image of SPD media policy—altered the arrangements surrounding the Dortmund pilot cable television project. Previously this project in NRW had been a powerful symbol of SPD opposition to private commercial television, being alone among the country's four new, experimental projects in debarring all private commercial satellite television channels, and in relying entirely on public-service programming. However, the new draft legislation envisaged an abrupt reversal of this 'fundamentalist' opposition to commercial television. Once again the SPD pragmatists seemed to have prevailed, in NRW as in Hamburg.

By early 1986 there had been not inconsiderable 'convergence' towards reconstructing a new minimum consensus, though there still remained strong pockets of resistance. The initially extremely difficult question of the legitimacy of commercial broadcasting in West Germany seemed to be nearly solved, yet a comprehensive *Staatsvertrag* still appeared to be elusive. As will be seen, this lack of complete agreement actually seemed to encourage a new pattern of Kleinstaaterei, or a pattern of uncooperative self-seeking for advantage by the individual *Länder*. Indeed, this pattern seemed to become more pronounced as it cut across the now weakening battle-lines left by the fierce ideological, party-political polarisation over the new media of the 1970s and early 1980s. However, in the meantime, commercial television, foreign and West German, had arrived. This simple fact alone meant that during 1986 considerable progress would have to be made in achieving a political solution and at very least the basis of a new regulatory framework for satellite broadcasting.

THE NEW SATELLITE TV CHANNELS

In addition to various foreign channels such as the British-based Sky Channel and Super Channel, by now received by West German cable systems, the development of new West German channels (or channels with heavy West German capital involvement) had been substantial since 1984. Nevertheless, developments were not rapid enough to fulfil the hopes of the West German Bundespostminister. Much of the channel capacity acquired by the Bundespost in accordance with Schwarz-Schilling's dream of a broadcasting 'revolution' long remained unsubscribed, at huge expense to the Bundespost itself. To some extent, this resulted from a wild overestimation of early demand for new channels. To some extent, it resulted from political problems, notably the failure of the *Länder* to agree a unified approach to the country's broadcasting future in such a way that would attract confident investment and stimulate positive action. As a result, the Bundespost endeavoured to give further encouragement to the new commercial sector, notably first of all by expanding its efforts to equip cable systems with advanced satellite reception technology and by liberalising arrangements for SMATV — more recently, by searching out additional 'low power' frequencies for off-air local channels — in large measure in order to relay the new satellite programmes in uncabled areas.

A limited compromise between the *Länder* resulted in agreement during 1984 on the allocation of two communications satellite television channels. One, the 'East-beam' of the ECS-1 communications satellite, was allocated to the public-service broadcasting organisation, ZDF, which produced a new channel, 3 SAT, in collaboration with Austrian and Swiss public-service broadcasters (the ORF and the SRG respectively). The bulk of its early programming originated in these public-service broadcasters' very sizeable libraries, although original material was also to be developed. 3 SAT ran no advertising, being financed entirely by its public-service operators. In addition to 3 SAT, ZDF also ran another new media venture, the pop-music channel Musik Kanal for West Germany's pilot cable projects.

The second beam of the ECS-1 satellite, the 'West-beam', was allocated to a consortium of publishers, in which the giant, right-wing Springer Press held a large share along with other major names in West German publishing such as Burda, Bauer and Holtzbrinck. Between them, with Bertelsmann and the WAZ group, these publishers dominated the West German newspaper and magazine market. This consortium produced a private commercial channel called SAT 1, which was franchised in the CDU *Land* Rheinland-Palatinate. It began commercial services at the beginning of 1985, feeding West Germany's (pre-existing and new) cable networks in the first instance, but aiming at a full DBS service in the future. A second commercial station also aimed at a rapid transfer to DBS, in order to capture a national West German audience. In January 1984, one full year earlier than SAT 1, the Luxembourg-based RTL Plus had started broadcasting terrestrially into the Saarland, and after August 1985 it shared the 'East-beam' of ECS-1 to transmit by satellite into West Germany. Through its offshoot company UFA Film-und-Fernsehen, the Bertelsmann group held a 40% stake in RTL-Plus and the West German WAZ newspaper group also took a 10% share in the channel. RTL Plus consistently captured higher ratings than its rival, SAT 1. Both of these new private commercial channels were totally dependent upon advertising for their revenue.

The new public-service channel, 3 SAT was centrally managed by ZDF. By the end of 1985, its signals fed cable systems in West Germany, Austria and Switzerland with a total potential audience of over one million households. Of the six hours of programmes daily (from 18.00 until midnight) around 65% were produced by ZDF, roughly 27% by ORF and about 8% by SRG. Although offering a mixed schedule of films, drama and news, the channel was deliberately conceived as a fairly 'up-market' culturally-oriented channel specifically for the German-speaking area of western and central Europe. 3 SAT's central conception bore clear recognition of the fact that the linguistic zones of Europe would increasingly define the various audience-zones of the future satellite age. By concentrating upon Germanic 'culture' and avoiding reliance on 'imported culture', the new channel aimed to strengthen 'home' production. However it was recognised from the outset that it would probably serve a minority, rather than a mass public; a programming philosophy which contrasted directly with that of the new private commercial operators.[21]

3 SAT was promoted very energetically by its public-service backers. The public-service broadcasters' desire now to compete with, and if possible strangle at birth, the new private commercial operators was barely hidden and not always sacrificed to 'high-brow' pretensions. Moreover, the new channel benefited greatly from the huge existing resources of its established public-service sponsors, both in programmes and personnel.[22] ZDF devoted only around 2-3% of its total expenditure (something in the region of DM 30-40 million per annum) directly to 3 SAT. During its first year of operation, nearly half of its programmes were 'repeats' of programmes already run on public-service channels; a

further 19% were 'live-transmissions' of programmes running simultaneously on the main channels. However, improvements were made during 1986, involving more material specifically made for the channel.

One highly significant point about 3 SAT is its implications for a 'cultural channel' for the Germanophone area of Europe stretching beyond the East-West border. The political and commercial attractiveness of such a channel receivable right across Eastern Europe, including the GDR, is obvious.[23]

In marked contrast to the public-service broadcasters, most of the new commercial operators — those organised in the SAT 1 consortium — had to start from scratch. Apart from their editorial expertise, already established through their publishing activity, their investment was of necessity new. Moreover, neither the SAT 1 consortium of new operators nor the more 'established' RTL-Plus would be able to recoup their costs until at least four million households were capable of taking their services (even this figure was an insufficient condition of success, in so far as the interest of the advertisers remained further dependent upon high scores in audience surveys as well as potential audiences). However, in 1985 SAT 1's advertising revenue remained disappointingly low at a mere DM 7.5 million, while its running costs were running comparatively high at DM 200 million (compared with Sky Channel'sestimated £12-15 million and RTL-Plus's estimated DM 60 million or approximately £16 million). Advertisers remained reluctant to take advantage of SAT 1 since by mid-1986 it could only access the nearly one million West German households already cabled, beyond which its appeal was limited largely to a German-speaking audience in Austria and Switzerland.

Like ZDF and 3 SAT, SAT 1 had its headquarters in Mainz. The SAT 1 news broadcasts, however, came live from Hamburg three times daily. One of these daily news bulletins, the 45 minute news at nine-thirty in the evening, was a particularly ambitious and fairly successful innovation; a conscious attempt to emulate the public-service broadcasters at one of their major functions. This news service was produced by a Hamburg-based company, called AKTUELL Presse-Fernsehen. Significantly, this company was itself a consortium of over 160 daily newspapers, in which the right-wing Springer Press held a 30% share!

Apart from this news service, the programming of SAT 1 remained predominantly entertainment-oriented. Although it included a very high proportion of 'international' (i.e. generally relatively cheap US origin) films and series, the West German publishers also appeared to be keen to produce substantial amounts of programming themselves. The programme schedule therefore also contained a marked bias towards 'homespun' popular magazine-, music-, quiz- and game-shows. The channel was accustomed to broadcast around eleven hours on weekdays and two additional hours on weekends. This keen ambition to quickly establish an expertise in production across a range of programme areas as well as merely packaging of 'bought-in' programmes was one of the main reasons for SAT 1's comparatively high early running costs.

SAT 1 was launched in a blaze of publicity and continued to benefit from a generous amount of press coverage afterwards. This was one of the advantages of having the publishers of many of the most widely-read magazines and newspapers in West Germany as backers. The publishing of detailed information concerning West German television (e.g. programme guides) had always tended to be monopolised by the very private commercial interests that were now moving into the new media and these all now gave an almost excessively high profile to SAT 1.

In the face of this onslaught, and albeit reluctantly, one by one the SPD-controlled *Länder* granted legal rights to receive the channel, during the period 1985-6. In the end, even the hardline SPD ruled *Land* of Hesse was forced to face faits accomplis.

The unique structure of the SAT 1 consortium very quickly brought its own problems, which in turn also help to account for the operation's comparatively high early costs. The channel's first year of operation revealed difficulties of cooperation and coordination between the various members of the consortium. Intra-organisational competition and duplication of effort proved to be rife, while business efficiency was undoubtedly impaired by the decentralised nature of the organisation with each component member independently responsible for its own budget, programme production and programme-buying.

During the summer of 1986 a major reorganisation of SAT 1 took place. There was both a streamlining and a strengthening of the share of the larger (politically conservative) publishing interests at the expense of the groupings of smaller participants within the consortium. Paradoxically, at first this reorganisation seemed to strengthen those components of the consortium which had produced the weakest performance during the early period of operation, to the disadvantage of those which had performed

127

surprisingly well. PKS alone had had a full year's headstart on the rest due to the valuable experience that it had gained in programming West Germany's cable pilot projects during 1984. AKTUELL had been, as noted, a surprising success from the start. On the other hand, problems had dogged Bauer, Burda, Springer and Holtzbrinck, who now appeared, perversely perhaps, to have increased their role.

Table 4.1: Shareholders in SAT 1

Name of the partner/shareholder	old	% Shares in SAT 1 July-86	Oct-86
Programmgesellschaft für Kabel-und Satellitenrundfunk (PKS)	40	35	40
Aktuell Presse Fernsehen (APF)	20	13	15
Springer	9.9	12	15
Burda	8.2	12	-
Bauer	6.1	12	-
Holtzbrinck	5.4	12	15
Neue Mediengesellschaft Ulm	1.4	1	1
Frankfurter Allgemeine Zeitung (FAZ)	1.4	1	-
Otto-Maier Verlag	1	1	1
Kabel Media Programmgesellschaft	6.6	1	-
'Special funds' (Sonderfonds)(on which Axel Springer and the APF each have an option of 5%)	-	-	13

(Sources: Medien Bulletin, July, 1986 p22; Media Perspektiven 6/86 pp 417-18; Media Perspektiven 10/86 p 686)

No sooner had this reorganisation occurred when the 'shake-out' continued and by October 1986, the publishers Burda, Bauer, FAZ and the Kabel Media Programmgesellschaft (KMP) had all pulled out, leaving the distribution of capital shares in SAT 1 once more highly favourable to the PKS. More significantly though, this second 'shake-out' concentrated even more control in the hands of the right-wing Springer Verlag (directly and indirectly, through the APF and the options on the Sonderfonds).

On the one hand, the reorganisation responded to a clear need for clearer decision-making structures and greater efficiency within SAT 1. On the other hand however, it also reflected the developing trend towards oligopolistic competition between giant publishers within the new industry (i.e. Springer versus Bertelsmann — a tendency which was not limited to West Germany.[24]

SAT 1 channel's private commercial rival, RTL-Plus benefited from a twelve month headstart. From January 1984 onwards it had broadcast direct terrestrially into the Saarland, Rhineland-Pfalz and parts of North-Rhine Westphalia. Far more importantly though, RTL-Plus was able to draw upon the vast accumulated experience, reputation and expertise in commercial broadcasting of RTL-Television in Luxembourg. Like the new public-service ventures, RTL-Plus was able to rely upon an established broadcaster's vast resources of talent and programming. During its first year of operation the amount of 'in-house' production material amounted to about 35-40%, rising to about 40-45% in 1986. Like SAT 1 though, the channel also relied heavily upon American 'soap-opera' drama and feature film material. However, as with SAT 1, advertisers did not exactly trip over each other to buy up spots on the channel, which rising from DM 1,300 per 30 seconds in 1985 to DM 2,500 were not inexpensive (compared to the DM 1,700 rate charged by SAT 1). Advertising revenue in 1985 amounted to a disappointing DM 18 million.

In mid-September 1986, the RTL Plus Deutschland Fernsehen GmbH & Co KG was established at Gutersloh, the West German home of Bertelsmann. The CLT now held 46.1% of the shares; Bertelsmann company Ufa-Film und Fernseh GmbH held 39.9%; the WAZ group held 10%; and the Deutsche Bank were the trustees of 4% of the shares. The obvious purpose of this (partial) 'homecoming' was to strengthen the channel's candidacy for a DBS franchise and also for the allocation of new 'low power' terrestrial frequencies (more on which later).[25]

West Germany's third and last private commercial satellite channel so far — the German Music Box — also started operations during 1985 using one of the Intelsat V transponders made available by the Bundespost. However, difficulties in finding enough advertising revenue led to the early pull-out of its original major backers, the Verlag M Dumont Schauberg publishers, and a major reorganisation of the company leaving it in the hands of a single individual, Wolfgang Fischer, with close political links with the Conservatives, based in Munich. Nevertheless, during 1986 it continued to survive.

That year also saw a marked revival of confidence and a determined response on the part of the public-service broadcasters to the new competitive challenge. The development during 1986 of a second public-service satellite television channel by the ARD, 'Eins Plus', using a transponder on the Intelsat-V F12 communications satellite, actually produced a measure of renewed political polarisation at the same time as it faced SAT 1 and RTL-Plus with additional competition from already established broadcasters. Eins Plus (like 3 SAT yet another co-operative venture with the Swiss channel SRG) opened on March 29 1986. At first it appeared that only the SPD *Länder* of Hamburg, Hesse, North-Rhine Westphalia, Bremen and the Saarland were going to accept it. Most of the CDU/CSU *Länder* seemed reluctant to accept it, since they were inclined to see it as yet another unwelcome attempt to take the wind out of the sails of the new commercial broadcasters before they had even begun to establish themselves upon anything near approaching a sound financial basis. CDU-governed Baden Württemberg and CSU-governed Bavaria even went to court to try to obtain a ban on the channel. They disputed the legality of 'Eins Plus' in view of the continued absence of an inter-state *Staatsvertrag* in which special provision for such a channel was made.[26]

As indicated earlier, this fragmented and precarious situation was further complicated during 1986 by manifest Kleinstaaterei or 'self-seeking rivalry' among the *Länder*. Thus, Bavaria's CSU (conservative) political masters aimed at using another of the Intelsat V transponders rented by the Bundespost to broadcast Bavaria's public-service 'Third Channel' — Bayern Drei — nationally across West Germany. This elicited a prompt response from the SPD North-Rhine Westphalia. Not to be upstaged, the SPD policy-makers in West Germany's most populous *Land* immediately applied to the Bundespost for a transponder so that their regional public-service broadcasting house might also 'go national'! Thus, the Federal Republic was faced suddenly with the scenario of an interesting ideological 'star wars' of satellite broadcasting. At the same time, the new private operators were faced with even more unwelcome competition from established broadcasting organisations. Revealingly, Bavaria's independent actions were widely seen by West German conservatives beyond that state's borders as amounting to a virtual 'betrayal' of 'official' CDU/CSU media-policy; the prompt reaction of North-Rhine Westphalia, it was feared, signalled the unleashing of a new dynamic that fundamentally destabilised an already highly unsatisfactory state of affairs. Bavaria had dangerously and irresponsibly 'broken ranks' by thus engaging in such an obvious case of *Kleinstaaterei!*

'RUMP' *STAATSVERTRÄGE* AND 'PARTIAL' *STAATSVERTRÄGE*

As the launch-date for the West German DBS satellite, TV SAT, approached during 1986, there was still no agreement upon which channels should be carried. Nevertheless, there seemed to be four strong contenders for four available TV channels — furthermore, a 'balance' was promised between two private commercial channels, RTL Plus and SAT 1, and two public-service channels, 3 SAT and Eins Plus. Therefore, there were good grounds for expecting the problem to resolve itself by default of excess competition for the channels. Moreover, in late 1986, the Federal Constitutional Court appeared to endorse the general direction of deregulatory new media policy-making led by CDU/CSU *Länder*.[27]

1986 saw the signing of two Staatsverträge : one for the south, and one for the north of the Federal Republic. The 'northern' one was signed by the CDU *Länder* of W Berlin, Lower Saxony and Schleswig-Holstein in March 1986. It was entitled 'state-treaty on the organisation of television by broadcasting satellite' and referred to the franchising of one television channel on TV SAT. The 'southern' one was signed between the CDU *Länder* of Baden-Wurttemberg and Rhineland-Palatinate, and the CSU *Land* of Bavaria in May 1986. It was similarly entitled 'state-treaty on the joint-use of one television and one radio channel on broadcasting satellites'. Both of these 'rump' *Staatsverträge* sought to establish faits accomplis by creating 'external' regulatory bodies, to ensure 'external pluralism', and by specifying joint franchising procedures by these respective bodies for two of the new television channels on TV SAT. These channels were clearly destined for private, commercial operators and there was not much guessing who these were going to be. The preamble of the 'northern' treaty made explicit the frustration of the CDU/CSU *Länder* at the failure to produce a comprehensive *Staatsvertrag* as a result of SPD tactics. It also made explicit that the signatories were not going to allow this to prevent them any longer from creating a regulatory framework

131

that was appropriate 'in view of the rapid national and international developments of transmission technologies'. The effect of this new development was rapid.[28]

The underlying economic and political pressures on SPD *Land* politicians towards pragmatic compromise that have already been described now led Hamburg to actually accede to the northern CDU 'rump' *Staatsvertrag*. As a result, the other SPD *Länder* were compelled to participate fully in the prompt drafting in June 1986 of a 'partial state-treaty for the reorganisation of broadcasting'. This envisaged that the first three of the TV SAT television channels should be allocated to the *Länder* and disposed according to separate 'state-treaties' of individual states. Presumably, it foresaw a possible third, SPD 'rump' *Staatsvertrag*, awarding a franchise to the public-service broadcasters, to complete the balance — since the fourth television channel was to be allocated to the ARD and ZDF together. The fifth channel was to be used for 15 stereo radio channels and two mono radio channels, to be divided (unequally, according to size) among the *Länder*. In practice, this amounted to a virtual restatement of the compromise position worked out two years earlier at Bremerhafen. 'Balance' would be ensured between two new public-service satellite channels on TV SAT and the two new private, commercial channels. At least, though, this reformulation seemed to introduce greater flexibility, in recognising the possible role of 'rump' *Staatsverträge* that were not intended to break up the ARD.[29]

However, there remained substantial political uncertainties. The CDU *Länder* objected now to Eins Plus. Agreement between the *Länder* on the all-important details of the programming of the West German DBS satellite clearly remained a problem. Moreover, the *Länder* had still to agree detailed regulations for programming content and, importantly, for advertising on the new satellite channels. Therefore, despite some promising new developments, Schwarz-Schilling still remained reluctant to sign his name to the DM 400 million contract for TV SAT 2, the back-up satellite which would eventually complete the system. In fact, Schwarz-Schilling had become increasingly concerned at the disappointingly slow pace of his 'broadcasting revolution'.

WEST GERMANY OPENS THE DOOR TO SMATV

During the period 1984-86, the cable programme had rolled ahead much as planned; investment actually rose to DM 1,500 million in 1986. However, progress was still widely considered to be too slow — by the Bundespost, by the equipment suppliers and by the backers of commercial television. Therefore, in order to hasten further the arrival of the 'broadcasting revolution' regulations for satellite-reception (SMATV) were also liberalised.

After many advance indications — including a directive from Schwarz-Schilling to West German dish manufacturers specifying an element of relaxed technical standards and encouraging them to start production — in July 1985 the Bundespost finally relinquished its monopoly of satellite-reception. From now on MATV systems and private individuals would be permitted to install their own satellite-reception dishes. Almost eight million West German households — a potentially huge market (compared, for instance, with just over one million homes in the UK) — were connected to MATV systems.

Although this measure might be considered to have promised a marginally disincentive effect on the demand for cable (still disappointingly low with take-up rates only creeping up to around the 30% of homes-passed mark), the measure reflected Schwarz-Schilling's concern to give commercial television a prompt boost. He hoped that in turn this would feed back into cable demand in the medium and longer term. Besides, having encouraged the German publishers to take the plunge and make huge loss-leading investment in satellite channels, he felt a responsibility to do his utmost to maximise their audiences.

Furthermore, the liberalisation of SMATV also had profound implications for the political debate raging over regulatory policy for satellite television. In particular, it made life very difficult for politicians who continued to argue against the reception of private commercial satellite television. Whereas they might argue that their regulatory jurisdiction extended to programme-transmission within their *Länder* — and this meant re-transmission of satellite-signals into the Bundespost's cable systems — they would find it near impossible to regulate for the direct reception of satellite signals by private individuals or by community cable systems (MATV and CATV). To attempt to do so would offend against the constitutionally-enshrined freedom of West German citizens to inform themselves from any available source of information

(Article 5 of the West German Constitution, the 'Basic Law'). Moreover, there would be a distinct anomaly in a situation where citizens, accustomed to being able to receive 'off-air' broadcasts from the German Democratic Republic if they so wished, were actually banned from receiving commercial West German satellite television by means of their own equipment. This was a compelling argument and in the end a compromise was reached even in Hesse in a law permitting the 'retransmission of satellite channels' within the *Land*.[30]

In June 1986, once again to stimulate developments perceived to be dragging too much, a new federal programme was launched 'to improve the general conditions surrounding the private broadcasting market'. Accordingly, the Bundespost determined to equip the majority of cable networks as quickly as possible with the latest satellite reception equipment. The Bundespost resolved also to take measures to remove outstanding technical obstacles to satellite reception where they persisted (the so-called Sonderkanal problem, which meant that only about 40-50% of connected households could receive new satellite programmes by cable). Finally, the Bundespost increased the marketing budget for its own cable programme from DM 8.5 million in 1985 to DM 15 million in 1986.

GERMANY ADOPTS THE D2 MAC STANDARD

A conflict of interests became abundantly evident in the summer of 1985, when Schwarz-Schilling signed a controversial agreement with his French counterpart, Louis Mexandeau, to introduce a new television transmission standard for DBS, called D2 MAC. Already accepted by the European Broadcasting Union (EBU), the new standard promised to be a much needed step towards European integration in the field of broadcasting regulation. It also promised much technologically, in particular the possibility of High Definition TV and also the enticing possibility of television programmes accompanied by simultaneous soundtracks in four different languages (containing great promise for a further degree of expansion of the European broadcasting market).

However, this decision amounted to a major setback for the recently established West German commercial satellite television operators, at least in the short run. They had constructed their plans on the supposition that they would be able to make an early, quick and profitable transfer to DBS in order to reach a much larger potential audience than the one million West German cabled homes. Adopting the D2 MAC standard quite simply meant the consumer would be unable to receive the transmissions without investing either in a completely new television set or a decoder-unit for his existing television set. Both the inevitable delay before industry could begin to supply this new equipment *en masse* and the extra cost to the consumer involved at a time of low to zero growth in disposable consumer income meant that the immediate public demand for satellite television was almost certain to remain at a disturbingly low level.

The commercial television lobby was furious at this 'betrayal'.[31] Their displeasure was increased by the knowledge that almost everyone else (except the consumer) seemed to benefit from this measure. The public-service broadcasters were encouraged by a policy development that, however unintentionally, upset the forward progress of their commercial competitors; the consumer electronics industry was also naturally delighted with the development of new markets for television equipment; finally, the Bundespost emerged from the affair a clear 'winner'. By the simple expedient of planning to convert the D2 MAC signals into the old PAL standard for the consumer and then feeding the suitably reconverted DBS transmissions into its growing system of cable networks, the Bundespost could actually promote its own cable programme and preempt a significant degree of competition from DBS.

THE SOLUTION: NEW 'OFF-AIR' LOCAL CHANNELS TO RESCUE THE SATELLITE BROADCASTERS

However, in the late summer of 1986, Schwarz-Schilling intervened once again to attempt to resolve these outstanding contradictions of policy and to give the 'broadcasting revolution' a much needed further stimulus. First of all, an internal directive of the Bundespost now emphasised a new priority to be given to finding spare 'low power' frequencies for 'local' private commercial television in all West German cities with populations over 100,000. This new policy initiative was confirmed when Schwarz-Schilling gave permission for the rapid start of just one such 'local' private, commercial television channel in Munich, called Kanal 59. Soon, however, it was announced that another 65

West German towns and cities, with a total population of 16 million viewers, would have their own 'local' television stations. This new policy was as good as an open admission that cable developments continued to be highly disappointing, and that, as a matter of urgency, further encouragement had to be offered to the new private, commercial satellite television operators. Off-air retransmission of their programmes by the Bundespost into local networks would help bring them the audiences that they needed in order to become viable. The two new CDU/CSU, northern and southern 'rump' *Staatsverträge* (mentioned earlier and incorporating Hamburg) made explicit in their opening articles that DBS programmers (to be franchised according to arrangements also laid out in the treaties) should have access to the new 'low power' terrestrial frequencies. This latest measure itself raised a certain amount of controversy, since smaller would-be private commercial 'local' operators now suspected that they would not receive much access to the new 'low power frequencies' which they legitimately feared would mainly benefit SAT 1 and RTL Plus.[32]

SATELLITE BROADCASTING IN WEST GERMANY: FUTURE PROSPECTS

Despite continuing disappointment at the end of 1986 it seemed highly unlikely that the leading new private commercial channels would collapse due to any further pull-outs of the major interests behind them: the 'shake-out' process seemed to have nearly completed itself. Currently, in order to stake their claim to a new field of enterprise the new channels' few giant backers seemed prepared to accept losses in the short and medium term — losses which they could easily afford. For example, the DM 250 million annual running costs of SAT 1 (in 1986) were but a fraction of the annual income of its most powerful backer; the Springer Press with DM 2,500 million per annum. Similarly, RTL Plus, costing rather less at around DM 60 million per annum (in 1986), was a mere fraction of the vast income of the giant Bertelsmann company, the world's largest multimedia giant, whose turnover climbed between 1984 and 1985 from DM 6,720 million to DM 8,000 million. The dynamic of unfolding competition between these mammoth backers seemed likely to become, in fact, the major factor militating against any further pull-out. After all, refusal to accept the challenge, by either party, would allow its competitor to make all the running in a field that still promised to be a money-spinner in the end. Indeed, during 1986 competition between the new satellite television channels for prime-time viewing showed signs of becoming increasingly fierce.

The pan-German market remained potentially very lucrative indeed, based upon a wide community of 'German' product and brand-names, a traditional 'German' trading sphere of activity in western and central Europe, and reinforced by the existence for several decades already of a pan-German television market. After all, one of the main reasons for the construction of cable systems in Austria and German-speaking Switzerland (and even in the GDR too) had been to relay West German television channels. In West Germany alone there was considerable pent-up demand for television advertising, which for example in 1983 had taken only 9.9 % of a total advertising expenditure of $5,226 million. This placed West German television advertising's percentage of total national adspend far behind that of countries such as Ireland (32.9%), Italy (32.0%), Austria (27.8%), the United Kingdom (25.1%), and France (15.5%); and also way behind the European average (16.3%).[33] One recent estimation of unsatisfied demand for television advertising in Western Europe suggested that West Germany represented, by a large margin, the richest pickings: the estimate of unsatisfied demand amounted to $356 million, followed by France ($278 million); in the United Kingdom, by contrast, it was estimated that there was no unsatisfied demand whatsoever in view of the ample opportunities to advertise on the ITV network, Channel 4 and TV AM. In West Germany currently at least, the limitation of advertising on the public-service channels to 20 minute blocks in early evening clearly meant that there was considerable potential for the commercial operators. The same estimation (mentioned above) suggested that a conservative assumption would be that in West Germany '...in an unconstrained world, 29% of existing adspend would go to television'.[34] Finally, as seen, the growth of a cable and satellite infrastructure continued to be actively promoted by the West German state.

During 1986, the entry of West Germany's large and well known consumer electronics giant Grundig into the satellite receiver market as well as heightened activity among the four specialist dish manufacturers — Kathrein Werke, Wisi, Fuba and Hirschmann — seemed to indicate that West German industry had begun to take seriously a growing opportunity. The take-off of the market for satellite reception equipment would be the clearest sign yet of the coming 'broadcasting revolution'. However, in May 1986 a heavy blow was dealt European DBS by the failure of the Ariane V 18 launch. Both TDF 1 and TV SAT had been scheduled for launch during 1986-7. This latest disaster followed several postponements of the schedule already due to technical and industrial factors and also one earlier failure of Ariane V 15 in September 1985 (with an ECS F3 communications satellite as payload). By the end of 1986, only optimists could predict a launch much before late 1987.[35]

In the meantime, however, the *Länder* agreed in March 1987 on a formula for a single, unified Staatsvertrag; it appeared that the SPD states had recognised the need to make substantial compromises. The *Ministerpräsidenten* envisaged a future broadcasting 'duopoly' with a private commercial sector co-existing alongside the public service sector. However, satellite channel capacity on the TV SAT satellites was allocated more in favour of the private sector; TV SAT 1 would carry the two public service satellite channels, Eins Plus and 3 SAT, and also the two new private satellite channels, RTL Plus and SAT 1. When TV SAT 2 eventually came into operation, making the TV SAT series fully operational, 3 SAT was to transfer to TV SAT 2 and a third channel would be awarded to the private sector and be carried by TV SAT 1. The German Music Box seemed to be a likely candidate for this third private commercial channel. All three channels would have local 'windows' since strictly speaking they were apportioned between the *Länder*.

For the private commercial operators an even more significant feature of the *Staatsvertrag* concerned its provisions for advertising. The treaty specified that any restrictions on television advertising on Sundays or on public holidays should be lifted, and that the private channels should also be allowed to run advertising after 20.00 hrs. (West German television regulations had previously only permitted television advertising between 17.30 and 20.00 and no more than twenty minutes per day.) In exchange for these changes, the SPD *Ministerpräsidenten* had won a firm commitment that the future prospects of the public service broadcasters would be 'guaranteed'; in this respect, though, the promise to raise the monthly television licence fee from DM 16.25 to DM 16.60 hardly seemed to be a generous *quid pro quo* by the CDU/CSU *Ministerpräsidenten*.

By late 1987, the prospects for the new private commercial operators were also considerably improving as a result of the pace of allocation of the terrestrial frequencies, most of which went to

SAT 1 and RTL Plus. Of these two channels, SAT 1 gained the lion's share obtaining major capacity for broadcasting in Munich, Mainz and Berlin. However, in order to win the competition for the highly desirable 'big city' terrestrial frequencies the major private commercial operators discovered the need to make considerable commitments to invest and locate production centres in these rival metropolitan centres. Nevertheless, at last the private sector was beginning to establish a wider audience.

CONCLUDING REMARKS ON THE POLICY PROCESS

The West German DBS initiative started life as industry/technology policy with firm roots clearly in that sector. As the implications for broadcasting policy (cultural policy) became clearer, the policy process became intensely complicated. In no small measure, this was due to the federal nature of broadcasting regulation in West Germany and political polarisation over the issue of commercial broadcasting. However, increasingly broadcasting policy makers were constrained to take into account the wider economic issues raised by the new technology.

Because of its federal nature, West Germany is a particularly useful case-study of how regulatory policies for the new media — under the impact of transborder satellite transmission — have fallen prey to the economic pressures on governments to 'promote' their regions and cities as 'media havens' in the age of 'open skies' broadcasting. Traditional political values and commitments to the wider cultural dimension of media policy seemed to have become appendages to this kind of more hard-nosed calculation of interest. Moreover, the highly polarised ideological battle-lines appeared to have been reshaped by the highly pragmatic battle-lines of regional competition.

For the SPD, the 'new media' have meant coming to terms with new and harsh 'realities', while for the CDU/CSU cable and satellitehave meant pioneering an exciting 'broadcasting revolution'. For the CDU/CSU the 'compulsions' (or 'structural constraints') of the international economy were largely unproblematic. Deregulation was in conformity with the political aims of their broadcasting policy. By contrast, the SPD policy makers were faced with an uncomfortable role of brokerage between economic/industrial policy and cultural/broadcasting policy aims. Tensions soon appeared between 'technocratic modernisers' and politicians in regional governments sensitive to the pressures of Standortpolitik or policy-making to favour location of media industries (inward investment), on the one hand, and a vocal alliance of 'media politicians' (like Albrecht Müller) and the grass-roots Left and the 'Greens', on the other. For a long time after the party's official policy reorientation, the SPD pragmatists continued to be embarrassed by the evident lack of enthusiasm for their readiness to compromise from the trade unions in particular, but also from some public-service broadcasters. Yet over the period of time under review, polarisation over media policy transferred progressively from the party political arena to the intra-party arena.

At the same time, it is interesting to note how Bundespost Minister Schwarz-Schilling has orchestrated developments from the centre. The convergence of previously discrete policy sectors — telecommunications and broadcasting — has enabled him to steer developments to the advantage of CDU/CSU new media policy, producing a wholly new element of federal influence on broadcasting policy. This represents a certain 'seepage' to Bonn of one of the few remaining areas of real *Länd* sovereignty. At the same time though, the responses at *Länd* level, from policy makers and broadcasters, over a period of time indicate the continuing typically decentralised nature of the West German broadcasting landscape. In particular, the prospect of regional satellite channels, with a distinct cultural accent and quite possibly an ideological bias as well, indicates the complex, and highly political, salience of broadcasting policies in a country which can be seen as an important microcosm of wider European developments.

REFERENCES

1. Finke, W. 'Das deutsch-französische Fernsehrundfunk-satelliten-projekt' in Kaiser, W. and Lohmar, U. *Kommunikation über Satelliten*. Springer Verlag, Berlin, 1981, pp. 150-63. On WARC 77 see: International Telecommunications Union (1977) *Final Acts of the World Administrative Radio Conference for the Planning of the Broadcasting-Satellite Service*. Geneva: ITU RE III/1982. Finke was *Ministerialdirektor* in the BMFT and responsible for TV SAT planning.

2. Luyken, G-M. *Direktempfangbare Rundfunksatelliten: Erklärung, Kritik und Alternativen zu einem 'neuen Medium'*, Campus Verlag, Frankfurt/Main., 1985, pp. 191-207.

3. Hauff, V. and Scharpf, F. *Modernisierung der Volkswirtschaft Technologiepolitik als Strukturpolitik*, Europaische Verlagsanstalt, Frankfurt am Main, 1975

4. Dyson, K. and Humphreys, P. 'Satellite Broadcasting Policies and the Question of Sovereignty in Western Europe', *Journal of Public Policy*, vol. 6, no. 1, 73-96. Also see Schmidbauer, M. *Satellitenfernsehen fur die Bundesrepublik Deutschland* Verlag Volker Spiess, Berlin W, 1983, A neo-marxist account is also both interesting and informative: Holzer, H. 'Satellitenfernsehen mit TV SAT' in Holzer, H and Betz, K. *Totaler Bildschirm-Herrschaft? Staat, Kapital und Neue Medien* Pahl-Rugenstein Verlag, Köln, 1984.

5. *Abkommen uber die technisch-industrielle Zusammenarbeit auf dem Gebiet von Rundfunksatelliten* in *Media Perspektiven* vol. 5, 1980, p. 342.

6. Luyken, *op cit.*, pp. 198-200.

7. For example, see Müller, A. 'Wieviel Medien braucht der Mensch' in Lang, U. (Ed/Hrsg.) *Der Verkabelte Burger — Brauchen Wir die Neuen Medien?* Dreisam Verlag, Freiburg i. Br., 1981.

8. Presse und Informationsamt der Bundesregierung *Zur Medienpolitik der Bundesregierung. Dokumente und Materialen*, Bonn, 1981.

9. The following description of the 'post-war consensus' and 'departure from consensus' draws heavily on Dyson, K. and Humphreys, P. 1986 *op cit.* For a much more detailed account of the historical development of West German broadcasting see: Williams, A. *Broadcasting and Democracy in West Germany*, Bradford University Press, London and Crosby Lockwood Staples, 1976. For an interesting background account of the recent political polarisation over broadcasting see: Kleinsteuber, H. *Rundfunk-Politik. Der Kampf um die Macht über Hörfunk und Fernsehen — Analysen*, Leske Verlag, Opladen, 1982

10. *Das Rundfunkurteil ('1 Fernsehurteil') vom 28 Februar 1961* BVerfGE 12/205. In this key judgement the Federal Constitutional Court ruled against Chancellor Adenauer's design to introduce a second national TV channel that would be Bonn-controlled and commercial. The judgement seemed to confirm the 'public-service' monopoly of West German television. Similarly, *Das Rundfunkurteil ('2 Fernsehurteil') vom 27 Juli 1971* BVerfGE 31/314. Much has been written on the special role of 'law in politics' in the Federal Republic. A very useful explanation is contained in G Smith *Democracy in Western Germany* Heinemann, London, 1979 pp. 185-94.

11. *Das Rundfunkurteil ('1 Fernsehurteil') vom 28 Februar 1961 op cit*

12. Kleinsteuber, H. *Rundfunk-Politik: Der Kampf um die Macht über Hörfunk und Fernsehen — Analysen*. Leske Verlag, Opladen, 1982. Williams, A. *Broadcasting and Democracy in West Germany*. Bradford University Press, London, and Crosby Lockwood Staples, 1976. More narrowly, detailed texts of the most important broadcasting laws (apart from the recent spate of legislation) are contained in *Broadcasting Laws* in the 'Documents on Politics and Society in the FRG' Series, published and distributed by INTERNATIONES Bonn-Bad Godesberg, 1979.

13. CDU *Freiheitliche Medienpolitik; 10 Thesen, Diskussionsgrundlage für den Medientag der CDU/CSU*. CDU, Bonn, 1978. CDU/CSU *Medien von morgen; für mehr Bürgerfreiheit und Meinungsvielfalt*. CDU, Bonn, 1984. FDP *Liberale Leitlinien zur Medienpolitik*. FDP, Bonn, 1979.

14. Hoffmann-Riem, W. 'Tendenzen der Kommerzialisierung im Rundfunksystem', *Rundfunk und Fernsehen* 32 Jahrgang, 1984/1, p. 36.

15. SPD *Leitlinien zur Zukunftsentwicklung der elektronischen Medien.* SPD, Bonn. SPD *Neue Medien; Aktionsprogramm der SPD zu den neuen Techniken im Medienbereich,* SPD, Bonn. SPD *Neue Medien und neue Techniken,* SPD, Bonn, 1982.

16. *Das Rundfunkurteil ('FRAG-Urteil')* vom *Juni 1981* BVerfGE 57/295.

17. Dyson, K. and Humphreys, P. *op cit.*

18. Konrad, W. in *Satellitenrundfunk — Medium der Zukunft?* Documentation of the Third International Media Conference of 18-20 November 1984 in Luxembourg, Hans Seidel Stiftung, Munich, 1984.

19. SPD *Medienpolitik: Eingeschränkte Öffnung für private Veranstalter; Beschlüsse des Essener Parteitages der SPD zur Medienpolitik.* SPD, Bonn, 1984.

20. A very detailed synopsis of the content of broadcasting regulations operative in the Federal Republic by the end of July 1986 (*Länd* media laws and *Staatsverträge,* and drafts) is contained in *Media Perspektiven: Dokumentation* 111/1986 pp. 125-211. Also see *Media Perspektiven: Documentation* 1, 11 and 1V/1986, for complete texts of individual laws, draft-laws etc.

21. Konrad, W. Ein kulturell akzentuiertes Programm des deutschen Sprachraums: 3 SAT. In *Media Perspektiven* 12/85, 1985, pp. 874-878.

22. *Ibid.*

23. *Cable and Satellite Europe,* December 1984, p. 46.

24. *Medien Bulletin,* July 1986, p. 22; *Media Perspektiven* 6/86 pp. 417-8; *Media Perspektiven* 10/86 p. 686.

25. *Media Perspektiven* 9/86 p. 608.

26. On 'Eins Plus' see Dietrich Schwarzkopf 'Eins Plus und Europa Television' *Media Perspektiven* 2/1986 pp. 74-80. On the reaction of a prominent CDU *Ministerprasident* and 'media specialist', Lothar Späth of Baden-Württemberg, see *Media Perspektiven* 2/1986 pp. 120-1.

27. *Der Spiegel* 46/1986 pp. 77-80. In keeping with the role of 'law in politics' in the Federal Republic, as well as with the characteristic 'legalistic style of opposition' in West German politics, the SPD had taken the CDU-inspired Lower Saxony media law of 1984 to the Federal Constitutional Court. The intention was to obtain a major ruling that would constrain the activities of the new private, commercial operators and that would delimit the competitive deregulation by CDU/CSU *Länder.* However, the Court appeared to give encouragement to the general direction of new media legislation, while at the same time ruling against certain features of the Lower Saxony law in question.

28. *Media Perspektiven: Dokumentation* 1/1986 pp. 43-56. For complete texts of both of these 'rump' *Staatsverträge.*

29. *Media Perspektiven: Dokumentation* 2/1986 pp. 110-4. For complete text of this draft 'partial' *Staatsverträge.*

30. Holger Börner *op cit.* pp. 113-21.

31. For the specific demands of this lobby see: Bundesverband Kabel und Satellitee. V. *Standpünkte und Forderungen* (Bonn: July, 1985).

32. On the first reactions of the smaller operators in Bavaria see Sissy Pitzer 'Munchner Anbieter gegen den Rest der Welt' *Medien Bulletin* (Frankfurt am Main) 7/1986 pp. 20-1.

33. Euromonitor Publications Ltd *Advertising in Western Europe* Euromonitor Publications, London, 1984. Quoted in: Tydeman, J. and Kelm, E J. *New Media in Europe: Satellites, Cable, VCRs, Videotex* McGraw-Hill, Maidenhead, UK, pp. 49-67.

34. Tydeman, J. and Kelm, E. J. *ibid.* 1986, pp. 64-5. The huge growth potential of the West German (and French) advertising market stands confirmed in another very recent cross-national Western European survey *The European Advertising and Media Forecast*, NTC Publications, London, 1985, prepared for the European Advertising Tripartite, Brussels.

35. TV SAT was launched in November 1987.

NB: This chapter is based upon extensive interviews and 'on the ground' research conducted by the author during the winters of 1984 and 1985 in Bonn, Hamburg, Mainz and Munich. The author would like to acknowledge that the article is based upon ESRC funded research into the new media, which was conducted in the period 1984-6 under the direction of Professor K Dyson in the School of European Studies at Bradford University (Reference Number E002/2099). The author would also like to acknowledge that many of the ideas contained in this chapter are the result of this joint research.

5

Satellite Broadcasting in the UK

André Goodfriend

BACKGROUND

For over 50 years the UK has been one of the major forces in world television. Its public service television structure has served as the model for numerous national television companies. The high quality of television programmes produced by British broadcasters has made Britain a major programme exporter, second only to the United States.[1]

Within the UK, television is strongly influenced by a public service tradition. Although it is often said that the public service principles governing British broadcasting emerged originally from the need to direct a scarce commodity to serve the common good, the increasing number of television channels has brought other arguments to bear on the need to preserve television's public service role. The basis for the existing public service structure has been described as follows: Broadcasting organisations must be:
   - independent of the government of the day;
   - independent of particular interest groups (political, religious, commercial or any other);
   - directly accountable to the viewing and listening public;
   - required to compete with one another;
   - assured of enough income; and so ... (be)
   - ...willing and able to sustain a domestic production base adequate to the provision of services which will satisfy the parliamentary expectation of them, and to assert national cultural values.
The structure devised in the UK to meet these requirements created two broadcasting authorities, each a public corporation independent operationally of the government of the day, and

accountable to Parliament for the provision of services which sufficiently match their expectation of them; and the allocation to each of them of an exclusive source of finance.[2]

The British Broadcasting Corporation (BBC), Britain's original public service broadcaster, is responsible for two television channels, both of which are funded by annual licence fees and prohibited from carrying advertising.

The Independent Broadcasting Authority (IBA) is responsible for regulating the remaining two terrestrial television channels ITV and Channel 4. Only ITV is operated as a commercial enterprise, funded through the sale of advertising. Although the IBA's second channel, Channel 4, broadcasts commercials, it is not reliant upon advertising revenue for its budget. Each broadcasting company affiliated with the IBA is required to contribute a percentage of its revenue to the operation of Channel 4, thus guaranteeing Channel 4 a budget, regardless of its viewing figures. In return for this guaranteed budget, Channel 4 broadcasts commercials sold by the ITV companies. This method of financing allows Channel 4 to fulfil its special public service charter of catering for the wide range of minority interests not normally covered by the major broadcasters.

All four channels are obliged to carry a certain amount of public service programming such as news and documentaries. The BBC is required to reserve time for the government and opposition political parties to present Party Political Broadcasts and Ministerial Addresses. These broadcasts are voluntarily carried by ITV.

This already intricate structure is now going through a period of uncertainty and change. New media are rapidly allying with or challenging the old, and there is no stable configuration in sight until the early part of the next century. The new technologies of broadband cable and satellite have played a fundamental role in leading UK broadcasters to increased competition, deregulation and change in the funding structure. As a result, new guidelines are having to be developed for new suppliers reaching new markets.

CABLE TELEVISION

Cable television has existed in Britain since 1951. Even at the outset cable systems were seen as a threat to the existing broadcast structure. As a safeguard to broadcast television, cable systems were restricted to relaying only those BBC and ITV broadcast signals which were available locally. Even the 'importation' of

British television signals from non-adjacent commercial television regions was not permitted. Cable systems languished in this restricted environment and soon became technically outdated. In 1972, twenty years after these cable systems were brought into operation, 60 per cent were still based on high-frequency multi-pair cable, while only 40 per cent made use of more advanced coaxial cable. Neither of these 'narrow-band' systems was capable of offering more than about six television channels.

Developments in information technology (IT) in the seventies and eighties had, however, come to the attention of successive Labour and Conservative governments. The Thatcher government's interest in IT culminated in the 1982 Information Technology Advisory Panel (ITAP) report, *Cable Systems*, which encouraged a speedy development of the new communications media. The ITAP report made five major points:

1 that cable systems be privately funded;
2 that existing cable programming restrictions be lifted;
3 that the need for a statutory body to oversee the cable industry be considered;
4 that technologically advanced cable systems be used;
5 that an early start be given to Direct Broadcast by Satellite (DBS) in order to assist cable systems.[3]

The Conservative government acted upon many of the ITAP recommendations. It created the Cable Authority and in due course awarded a number of 'broadband' cable franchises. Eleven pilot franchises were awarded in 1983, prior to the setting up of the Authority. The Cable Authority has subsequently awarded another ten franchises bringing the total of UK broadband cable franchises to 21. By 1987, only 10 of these franchises were actually in operation. Broadband franchises are required to make use of advanced cable technology to provide an extremely versatile, interactive service.

At the same time that broadband systems were being licensed, existing 'narrowband' systems were released from the 'must carry' requirement to transmit the four broadcast channels provided that other means could be found to ensure their reception by cable subscribers. This was usually accomplished by supplying subscribers with television aerials, thus releasing the four to six channels of these 'upgraded' systems to carry other programming. Although some of the new cable television channels make use of locally originated productions, 'other' programming has in most cases meant satellite television.

THE DEVELOPMENT OF SATELLITE TELEVISION IN BRITAIN

Satellites have been used in connection with television in the UK ever since the Telstar experiment in 1962. The expense of the equipment involved, however, prohibited their widespread use. In the 1970s television signals made up only a small portion of European satellite transmissions. The dishes required to receive these signals were several metres in diameter, thus effectively limiting their television use to international news agencies and major broadcasters. Such uses continue to be of importance and TV organisations around the world still use satellite facilities for such things as conferences, interviews and events.

However, the late 1970s and the 1980s have seen a fundamental change in both the uses of satellites and the consumers of such services. In 1978, the European Space Agency launched the Orbital Test Satellite (OTS). Initially OTS was used for relaying telephony and other point to point communication. Then in 1982, the British based Sky Channel, using OTS, became the first European satellite television channel to be transmitted across European cable systems.

Similarly, the Eutelsat and Intelsat series of satellites, launched in the early 1980s were originally intended for telecommunications rather than cable television, but the growth of interest in cable systems had greatly increased the demand for satellite delivered television and the satellites have adapted their service accordingly.

Apart from the increasing use of satellites to relay television signals, another change was taking place. Satellite programming was becoming 'broadcast' rather than 'distributed'. Satellite television in the 1970s and early 1980s was intended to be received by a terrestrial broadcaster who would then either edit it for re-broadcast or relay the transmission unchanged. As both satellite and reception technology have progressed, the emphasis began to shift to the creation of a DBS system capable of direct to home (DTH) reception, obviating the need for cable systems to carry the increased number of television channels.

As these developments have taken place, telecommunication satellites have become grouped into three major categories: low power, medium power and high power. These three stages of satellite development are not simple increases in transmission

147

power. Each one fulfils a slightly different function with the most obvious difference being between the medium power and the high power satellites.

Although television satellites in previous decades had transmission strengths as low as two watts, low power communications satellites today have transmission strengths of between 10 and 20 watts. The footprints of the European low power satellites cover, or at least touch, most western European countries. Yet the area where the signal strength is strongest, enabling DTH reception, is relatively small, usually consisting of one or two countries or portions of countries. Even within this beam centre, the satellite receiving dish must usually have a diameter of about 1.2 metres. Outside the beam centre the required dish size increases rapidly as the strength of the signal decreases.

Medium power satellites will have roughly double the transmission strength of low power satellites. Thus, a signal even stronger than that at the beam centre of the low power footprint will be received by most countries in Western Europe. Not only will the signal be stronger but the footprint of the medium power satellites has been shaped to fit the contours of the area being served. Low power satellites, with circular or oblong footprints, wasted much of their signal on the ocean.

While medium power satellites are seen as an upgrade of the low power satellites, high power satellites have been designed to serve a completely different function. They were designed specifically for DBS broadcasting within individual countries, as envisaged in the WARC '77 guidelines. High power transmitters over 200 watts were to be combined with a very narrow footprint, covering just one country, in order to provide an intense signal which could be received directly by the domestic viewer using reception equipment measuring under 90 centimetres in diameter.

THE CURRENT SCENE

As this is written, the UK is in the peak area of the footprint of two low power satellites — Intelsat V F-11 and Eutelsat's ECS F-1. Both these satellites deliver a wide range of television services to cable networks. Satellite TV is in many respects the life-blood of the British cable industry.

UK cable franchises relay to their subscribers all of the satellite-transmitted English language programmes except CNN. A

Table 5.1: The satellite services receivable in the UK with a 1.2 metre dish in 1987

Satellite Name:	ECS-F-1	Intelsat V F-11
Owner:	EUTELSAT	INTELSAT
Launch:	6/83	3/82
Operational:	1983	1983
Position:	13°E	27.5°W
Strength:	20 watts	10 watts
No. of TV Transponders:	10	3
Programmes:	3SAT (G)	Arts Channel (E)
	FilmNet (N)	Children's Channel (E)
	RAI Uno (I)	CNN (US)
	RTL Plus (L)	Lifestyle (E)
	SAT1 (G)	Premiere (E)
	Sky Channel (E)	Screen Sport (E)
	Super Channel (E)	
	Teleclub (S)	
	TV 5 (F)	
	Worldnet (US)	

(G)=German (I)=Italian (L)=Luxembourg (E)=English (F)=French (US)=US (S)=Swiss (N)=Netherlands

selection of foreign language broadcasts is also available on cable. The size of the TVROs used by the cable franchises enables them to relay not only the relatively strong Intelsat and Eutelsat signals, but weaker signals such as the Russian Gorizont service as well (Table 5.1).

The next major satellite broadcasting development to affect Britain will be the launch of several European medium and high power satellites. Medium power satellites are being used for the quasi-DBS, pan-European systems envisaged by the private, Luxembourg based consortium, SES/Astra and by the EBU backed telecommunications consortium, Eutelsat. National DBS ventures, which are scheduled to come into operation at the same time as the pan-European systems, will make use of high power satellites. Britain will be the third European nation — after Germany (SAT1) and France (TDF1) — to launch a high power system.

ANDRÉ GOODFRIEND

SATELLITE PROGRAMMING

Britain has already established itself as a strong presence in the creation of international satellite programming. About a quarter of the satellite channels currently available throughout Europe originate in the UK.

In mid-1987, the satellite channels originating in the UK were:

The Arts Channel, with its offices in Wales, transmits a mix of opera, jazz, drama, classical music, literature and the visual arts for three hours each day. Launched in 1985, the Arts Channel carries no advertisements and is financed by subscription fees and sponsorship. 84% of Arts Channel material comes from EEC countries while 12% originates in North America and 4% in Russia. Twenty percent of its programming is produced by the channel itself. Due to low revenue, the Arts Channel tries to keep its costs to a minimum. This is one reason why it has the awkward but relatively inexpensive satellite transmission time of 6:00am to 9:00am.

The Children's Channel, as its name indicates, produces programmes for children. It was begun in 1984 and currently transmits for eight hours a day, between 7:00am and 3:00pm. It derives its revenue primarily from advertising, broadcasting 48 minutes of commercials a day. Subscription fees are also a source of revenue. Only 30% of The Children's Channel's programming originates outside the UK, with 20% of its programmes produced in-house.

Lifestyle is a channel catering largely for women at home. It broadcasts programmes of general interest such as cookery, fitness, talk shows, advice, soap operas, book reviews, gardening, DIY. Lifestyle was launched in 1985 and derives its revenue from advertising. It broadcasts a maximum of 40 minutes of commercials a day. 75% of Lifestyle's programmes originate in the UK. It transmits for 4 hours a day, between 9:00am and 1:00pm.

MTV (Europe) is the European counterpart of the American 24-hour pop music channel of the same name. The format has been altered only slightly for the European market with a greater emphasis on music from the European charts. Having taken on two major British investors, Mirror Group Newspapers and British Telecom, MTV began its service in 1987 and hopes to succeed in spite of the commercial failure of Music Box.

Premiere shows recently released English-language films. It was formed in 1984 by the American film channels and studios HBO,

Showtime, Columbia Pictures and 20th Century Fox, along with the British Thorn EMI and Goldcrest Studios. Thorn EMI and Goldcrest later dropped out and their place was taken by Robert Maxwell's Mirror Group Newspapers. Premiere carries no advertising and receives its revenue from subscription fees. It transmits for 10 hours a day beginning at 3:00pm.

Screen Sport transmits 7 hours a day of sports news and events, beginning at 5:00pm. It was launched in 1985 along with Lifestyle. Screen Sport's revenue comes primarily from subscription, although it will occasionally broadcast up to 40 minutes a day of advertising. Sponsorship of programmes is also a source of income.

Sky Channel is a family entertainment channel transmitting a variety of drama, action series, comedy, sports, movies, documentaries, stereo pop music and teletext. Sky Channel was the first European satellite channel, having begun transmission in 1982. It derives its revenue solely from advertising and transmits 144 minutes of commercials a day. Sky Channel's programming comes from both European and American sources. The high percentage of American programmes in its schedule has given it numerous critics. In response to this, it has begun broadcasting in-house productions, which now form as much as 46% of its weekly schedule. Due to a lack of agreement with the television actors union, Equity, few British productions are broadcast.

Super Channel/Music Box was created in 1986 as a channel which would show the 'best of British' general entertainment programmes. Both the BBC, which is the Super Channel's main source of programmes, and ITV have agreed to provide programmes to Super Channel first before making them available to other satellite channels in the same market. Super Channel's format makes it a direct competitor to Sky Channel. Because of Super Channel's high British programme content, the battle between Super Channel and Sky Channel is often characterised as being between British and American programmes. Prior to Super Channel's existence, Music Box existed as a 24 hour music video and concert channel. Super Channel incorporated a shortened version of Music Box into its own schedule and took over its satellite transponder. Independent Television News (ITN) which produces the television news for the ITV companies, produces a special international evening news bulletin for Super Channel. As with Sky Channel, Super Channel's revenue comes from advertising. It also transmits 144 minutes of commercials a day.

151

Super Channel is financially supported by every ITV company except Thames TV, the largest of the independent broadcasters. Thames, while providing programmes for Super Channel, has chosen to invest in other new media ventures, such as The Children's Channel and the SES satellite venture, Astra.

BBC radio and television services

In May 1987 the BBC began using Intelsat V to transmit its terrestrial channels to Denmark. This BBC service was not created especially for satellite; it is a satellite relay of its normal British service. The satellite transmission was initiated at the request of the Danish cable system. According to reports, Sweden, Norway and West Germany are also interested in receiving this satellite transmission for their cable systems. France, the Netherlands and Belgium already receive a strong enough BBC signal from terrestrial transmitters to include BBC programming on their cable systems.

At the same time that the BBC began transmitting its television programming on Intelsat the BBC began using the two mono channels which came with the Intelsat transponder to transmit the BBC World Service radio station. The decision to use satellite to transmit the radio service was due to the poor quality of short wave radio reception. In order to increase the reception of the World Service even further, the BBC has begun transmitting on Eutelsat as well using the mono channels available on the transponder leased by Super Channel. Super Channel agreed to the transmission of the World Service on their transponder in the belief that it would add prestige to their service and act as an extra incentive for satellite listeners and viewers to tune to Super Channel.

The British satellite broadcasters are not public service entities. All are commercial ventures aspiring to profit. In many ways this contrasts with previous British broadcasting tradition and with existing European attitudes. France, Italy, West Germany, Sweden and Norway all transmit public service channels by satellite. Satellite television has ignited the debate over the necessity and future of public service broadcasting. If the Arts Channel, with its 'high-brow' content, can survive without public funding, why is it necessary for the State to fund the public service broadcaster? Super Channel, which is made up of private investors, provides a

Table 5.2: Summary of UK satellite channel data

Channel	Began Service	Min/Day of Commercials	Hours/Day of Service
Arts Channel	1985	No	3
Children's Chan.	1984	48	8
Lifestyle	1985	40	4
MTV	1987	n/a	24
Premiere	1984	No	10
Screen Sport	1985	40	7
Sky Channel	1982	144	18
Super Channel	1986	144	20

varied schedule and its own news service, do public broadcasters provide more?

The other side of the argument is that profitability continues to elude the satellite broadcaster, thus limiting the amount of original and 'worthy' material that these broadcasters are able to present. Sky Channel, one of the most mass oriented channels, had hoped to reach the break-even point by 1986, but may have to wait until 1988/89 before it comes into the black. This current unprofitability will undoubtedly cause the above configuration of channels to alter as some broadcasters are no longer able to maintain their budget and others find new opportunities opening in a rapidly growing market. It remains to be seen whether 'high-brow', niche market channels, such as the Arts Channel, will be able to survive against mass appeal channels such as Sky Channel and Super Channel.

Yet even in a changing market, it is likely that British programmers will continue to dominate the European satellite scene. Britain currently has the most satellite channels of any one country and its two mass appeal channels Sky and Super Channel are the most popular satellite channels in Europe. New British channels are being added each year while additional new media formats and markets are actively being developed.

The popularity of Britain's satellite channels has enabled them to secure transmission on the cable systems of countries throughout Europe. Even Japan's new DBS venture relays the Super Channel news as well as a selection of BBC and ITV programmes.

The extent to which British based satellite programmers are being relayed on the cable systems in other European countries can be seen in Table 5.3.

ANDRÉ GOODFRIEND

INVESTMENT IN SATELLITE TELEVISION

As indicated earlier, the new media in Britain are characterised by their strong private sector base. This dominance of private sector over public sector financing stems from the Conservative government's policy of encouraging free market competition within industry. Given the opportunities presented by the advent of new media technology, even television, with its strong public service roots, became a suitable candidate for privatisation. Super Channel, which is the publicly funded BBC's main new media market, raised its funds through venture capital investment. This new grey area where public and private sector combine to form one entity has challenged existing frames of reference. Although most of Super Channel's investors are IBA licensed broadcasters, the presence of the private company, Virgin Vision, was grounds enough to grant Super Channel only associate member status in the EBU.

Numerous other companies, not wanting to be left out of the new media environment have invested in several satellite channels as well as cable franchises. British Telecom is notable for its high investment in every aspect of the industry, programming, cable and satellite distribution. Recently privatised, BT is making sure it can compete on every front of the communications landscape.

The two media giants Rupert Murdoch and Robert Maxwell, both coming from the publishing world, are also carving out new media spheres of influence which span several countries. Although it is arguable which nation Murdoch represents when speaking of world media, Sky Channel is a British based company which is now carried on the cable systems of 19 countries.

Robert Maxwell, the head of Mirror Group Newspapers, has concentrated his efforts on establishing a strong European media base. Within the UK Maxwell controls most of the 'narrow-band' British cable systems. Apart from his current two satellite television channels, Maxwell has, through the Pergamon Media Trust and Maxwell Media, a large stake in the Bouygues-led consortium which recently took control of France's first national network, TF1. Another Maxwell company, Media International, is seeking one of the channels on the French DBS satellite, TDF-1.

Within the UK, satellite television is opening up what had been a tightly controlled system to a number of other players. The BBC and IBA become but two members of a burgeoning international club, some of whose members don't have television as their main

Table 5.3: European cable penetration of British satellite channels

	Arts Channel	Children's Channel	Lifestyle	Premiere	Screen Sport	Sky Channel	Super Channel
Austria						X	
Belgium		?				X	X
Denmark	?		?		?	X	X
Finland	X	X	?		X	X	X
France			?			X	X
Germany					?	X	X
Greece						X	
Hungary						X	X
Iceland						X	?
Ireland	X	?	X		X	X	X
Luxembourg						X	X
Netherlands	X	?				X	X
Norway	X		?		?	X	X
Portugal						X	?
Spain			?			X	X
Sweden	X	X	X		X	X	X
Switzerland	?					X	X
UK	X	X	X	X	X	X	X
Yugoslavia						X	?

X = currently distributed by cable
? = may soon be distributed by cable

interest. For the most part, however, those with the largest share of the new media pie are not complete outsiders to the communications field.

The major players come from publishing, book retailing, film, music and telecommunications. For many corporate investors, satellite television is the next logical step in advancing their own corporate goals. Newspapers give way to 24-hour televised news and entertainment services with a global reach. Music leads to

Table 5.4: Major investors in British satellite TV channels

	Arts Channel	Children's Channel	Life-style	MTV	Premiere	Sports Channel	Sky Channel	Super Channel
British Telecom		22%		24%	30%			
Central TV		22%						17.5%
Columbia					10%			
Commercial Union	28%							
DC Thompson		22%	15%				3%	
Equity & Law	28%							
Granada Group								23.2%
Home Box Office					10%			
John Griffiths	21%							
London Weekend TV								11.7%
Mirror Group				51%	30%			
News International							83.5%	
Showtime					10%			
TV South	7%		18.5%					11.7%
Thames TV		22%						
Thorn EMI		12%						
20th Century Fox					10%			
Viacom Int'l				25%				
Virgin Vision								15%
WH Smith	14%		48.7%			91%		
Yorkshire TV			18.5%					6.9%

televised music videos, while telephones and television become part of the same communications circuit.

The distinction between the various media — print, film, radio, TV, telephone — becomes blurred when we reach the stage of satellite television. Newcomers to television abound, but at closer inspection, the major players have been involved in some aspect of mass communication for many years.

Table 5.4 illustrates the cross-investments within current British satellite TV channels.

THE REGULATORY STRUCTURE

Originally the responsibility for broadcasting in Britain rested with the Post Office which also regulated telephone and telegraph communications. In this respect, the British regulatory structure was similar to that in many other European countries. In 1969 the Post Office was restructured and ceased to be a branch of the government. In 1974, the Department of Trade and Industry (DTI) took over responsibility for telecommunications and the Home Office became responsible for broadcasting.

A decade later, the introduction of cable and satellite, along with the deregulation of the British telecommunication system, has necessitated a further reconfiguring of the British media regulatory structure. While the Home Office remains responsible for terrestrial broadcasting, responsibility for cable and satellite broadcasting has been given to other agencies, each dealing with a different aspect of the new media.

The DTI in its role as the telecommunications authority, issues licences for telephone and cable franchises, SMATV systems and domestic TVROs. The current one-off licence fee for a domestic TVRO is £10.00. The DTI is also responsible for radio frequency allocations and management of the spectrum. To illustrate the similarity between cable television and telephone systems, the DTI licence for cable and telephone systems is almost identical.

In order to oversee the smooth privatisation and deregulation of the telecommunications industry the Office of Telecommunications (OFTEL) was established. While having no regulatory power of its own, OFTEL acts as a government watchdog, ensuring that telephone and cable franchises comply with their DTI issued licence and engage in fair competition.

Figure 5.1 The British new media regulatory structure

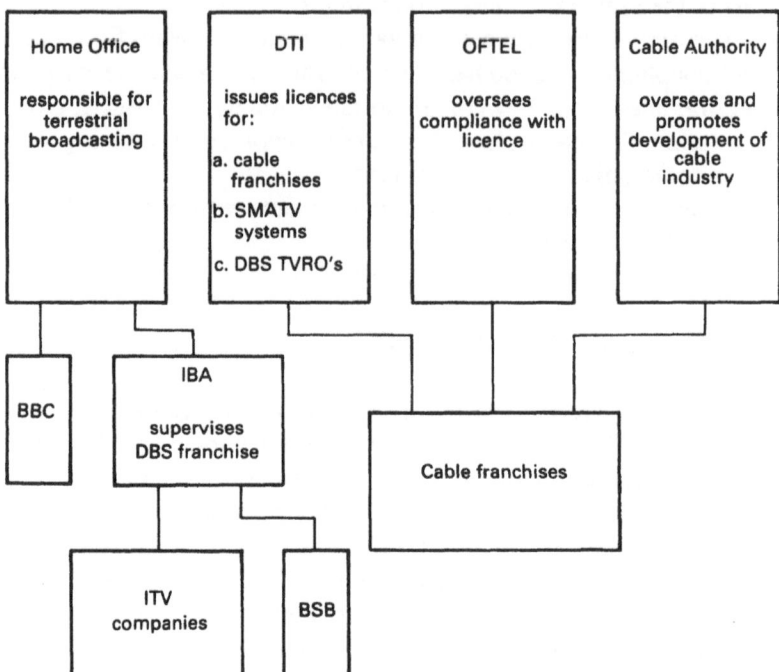

The Cable Authority, which was established in December 1984 following the enactment by Parliament of the Cable and Broadcasting Act, oversees and promotes the development of the cable industry. The Authority is responsible for awarding the cable franchise which is then licensed by the DTI. Whereas the DTI and OFTEL join in the supervision of the hardware aspects of cable, the Cable Authority is the primary monitoring agency with regards to cable programming.

The Home Office had the initial responsibility of designating the body responsible for DBS broadcasting. After many developments, described later in this chapter, authority for supervising DBS was given to the IBA, itself responsible to the Home Office. The IBA in turn awarded the DBS franchise to British Satellite Broadcasting (BSB) (Table 5.5).

Given the abundance of regulators and overseers, it is sometimes difficult to know who is or will be responsible for what (Figure 5.1). For example, although the IBA has been given the responsibility for overseeing the DBS franchise, the Cable Authority will still have to monitor the DBS programming since it will be carried by all of the

UK cable franchises. In light of this, it has been proposed that DBS be made the responsibility of the Cable Authority. Considering that the IBA only recently issued the DBS franchise, it is unlikely that this will happen in the near future.

Even though many satellite broadcasts reaching the UK originate in other countries, this maze of agencies is able to retain a certain degree of control over the content of programmes reaching the British television screen by regulating the major domestic means of programme distribution, i.e. the cable industry.

Although terrestrial television is not currently subject to the Obscene Publications Act, thus allowing the BBC and IBA to use their own judgement in broadcasting 'adult' programmes, cable television (and, by extension, satellite television) is subject to the Act. It is the Cable Authority which supervises cable television's compliance with these guidelines by responding to complaints about cable programming or services.

In this regard, channels which show films or programmes deemed to offend public taste and decency are not permitted on UK cable. For the moment, this restriction does not cause the Cable Authority much difficulty as there are no such satellite channels currently transmitting to the UK. Individual programmes do, however, occasionally transgress British guidelines. Such a case recently occurred when the Cable Authority banned a pop video, deemed obscene, from appearing on Super Channel/Music Box. In this instance, the difference in standards between terrestrial and cable broadcasters was made evident by the fact that the British radio stations chose not to ban the song, but only to schedule it in the late evening.

The Cable Authority's influence even stretches across national boundaries. In 1986 it complained to the French satellite channel, TV5, that a television programme sponsored by Renault was unsuitable because, according to Authority's guidelines, a programme 'should not display undue emphasis on the product, name or service of the sponsoring company'. The programme was, according to the Authority, virtually one long commercial for Renault cars. In 1987, the Authority requested that a Swiss documentary on prostitution, *Les Trottoirs de la Suisse Romande*, also transmitted on TV5 not be repeated as scheduled because it offended British standards of public taste and decency.

In order to ensure that the Cable Authority is able to review programmes against which complaints have been lodged, British based programmers are required to make their videotapes available

159

to the Cable Authority, the Broadcasting Complaints Commission or the Police. The Cable Authority requires the individual franchisees to maintain a copy of all foreign satellite programming which has been transmitted on any UK cable franchise during the previous 30 days in order to substantiate complaints made against foreign programmers.

The only satellite channel transmitted to the UK which was, for a time, prohibited from transmission on cable is the American based Cable News Network (CNN). The complications surrounding CNN's transmission are discussed later in this chapter.

THE CHANGING RELATIONSHIP BETWEEN CABLE AND SATELLITE

As direct-to-home (DTH) satellite broadcasting becomes a more important force in Britain's new media landscape, the ability of national agencies to regulate domestic reception of foreign television channels may be greatly diminished.

The cable industry in the UK has not grown as rapidly as had been predicted in the years immediately following the ITAP report. Conservative predictions made then suggested that by 1990 there would be about 8 million cable subscribers. Predictions now put the number of UK cable subscribers in 1990 at about 600,000. At the beginning of 1987 about 200,000 households in Britain were receiving satellite transmitted programmes via cable.

At present, DTH reception in the UK is insignificant when compared with the number of cable households. In June 1987, the number of households with a licensed TVRO was about 3,500. Yet although the number of domestic TVRO's in Britain is small at present, it is expected to grow rapidly within the next five years. By the early 1990s forecasts for the UK put the number of households with TVRO's well over the number of households receiving cable television.

SES Astra, in publicising its forthcoming medium power satellite, used figures from Communications and Information Technology (CIT) Research Ltd. to demonstrate the high potential demand for direct-to-home (DTH) satellite television.[4]

CIT uses three factors to determine satellite television's acceptance by the public: price, antenna size, and the number of programmes of interest available. If the dish size is 0.6 metres or below, if the price is $435 or below and if there are at least four channels of interest, the CIT 'model' asserts that the growth of satellite television in non-cabled areas will be equivalent to the growth of video cassette recorders.

While current low power satellites require a TVRO ranging in size from 90 cm to 2.2 metres, with prices ranging from about £600 to £1800 to receive the signal, prices and sizes for TVROs in the next five years fall well within the parameters of the CIT 'model'. If this 'model' is correct, Britain, with a penetration rate of well over 50% for VCRs, could be a very lucrative DTH market. SES Astra predicts that by 1996 61% of British homes will have a TVRO while less than 10% will have cable.

Given cable's sluggish UK performance and the high prospects for satellite, a UK-DBS is being looked at with renewed interest. DBS has been an on-again/off-again proposition in Britain since the late 1970s, even before the Government's push towards broad band cable in 1982. A UK-DBS now appears set to become reality before the end of the 1980s.

DBS — THE 10-YEAR STRUGGLE

DBS was originally envisaged as the ideal broadcasting system for public service television. One major principle of public service television is that broadcast programmes should be available to the whole population. While low power satellites gave universal coverage of Britain, the equipment required to receive the signal was large and expensive. DBS would give universal television coverage with receiving equipment that was small and affordable. The number of channels available on DBS would also allow it to satisfy another principle of public service broadcasting, that programmes cater for all interests and tastes.

Another consideration was that whereas low power satellites have a large footprint, covering several countries, DBS high power satellites use a very small beam focused on just one country. This would enable DBS broadcasts to cater to the public service interests of the receiving country in a way which transnational low power broadcasts would not be able to do.

In order to ensure that DBS would develop as a national broadcasting service, with minimal interference or overspill from adjoining systems, the World Administrative Conference (WARC) divided the 12 GHz band between nation-states at its 1977 meeting in Geneva. The WARC alloted the UK 5 DBS television channels, as it did with most other European States.

With the WARC '77 plan in operation, the British government began to consider developments in direct-to-home satellite broadcasting. In 1980 the Home Secretary announced the setting up of a study into the possible options for UK DBS. The Home Office report was published in May 1981 and established the framework for subsequent developments in policy.

Not developing DBS at all was never a serious consideration. Both the Home Office and the DTI were aware of a variety of economic, industrial and political pressures which favoured some degree of involvement in DBS. Apart from the generally held belief at the time that involvement in the new technologies, as represented in part by DBS, was the only way to remain industrially competitive in the hi-tech world of the future, the government was committed to those industries which had both internal and external industrial benefits. No start at all would have meant that the UK no longer considered itself a player in the European industrial landscape. It would have also left the UK open to the cultural invasion of foreign DBS transmitted programming against which the UK had no domestic alternative. In contrast, a UK DBS venture, under the control of the government chartered BBC could be developed in a way which would ensure some control over programme values and standards.

While it was taken for granted that a British DBS venture would be of immense benefit culturally, commercially and industrially, developing a full scale multi-channel DBS service for an early start was considered too radical a step just at a time when the terrestrial broadcasting structure was itself undergoing drastic changes. Channel 4 and the extension of television programming into the morning hours were just getting under way. Cable television was also attempting to woo its share of the television audience. It was thought that five new channels at this delicate stage would have endangered the whole existing structure and quality of British television.

Given these considerations, a compromise consisting of an early but limited start was seen as the most practical way to proceed. In May, 1982, the BBC was awarded the licence to run DBS in the UK; the proposed starting date for transmission was 1986. In order to achieve the government's commercial and industrial goals, specific conditions were attached to the BBC's management of the venture.

DBS in Britain was to begin with two channels transmitted from a satellite built by United Satellites (Unisat), a consortium made up of British Aerospace, Marconi and GEC, all British companies. The project was costed at about £24m per year for seven years. Although these conditions forced the BBC to spend more money than it would have had to were it able to select from the other, less expensive satellite systems on the market, the industrial goals were considered to be of overriding concern. This inability to consider cost effectiveness was a major point of contention throughout the existence of this first British DBS venture.

Apart from the government impositions with regard to hardware, it also set conditions with regards to the financing of the project. The BBC was prohibited from calling on public funds using existing sources of revenue to fund the project. Instead, it was given a supplementary charter which enabled it to borrow up to £225m to fund the project. The major source of funding for the DBS venture itself was to come from subscriptions.

Although the BBC had prepared detailed plans outlining the manner in which it would establish a British DBS service, numerous complications soon arose. Apart from the extremely high cost of the venture, which the BBC had underestimated, the government had begun to turn its attention to cable television. If a cable system could be developed, the benefits derived from DBS television would be greatly reduced. The ITAP report while encouraging both cable and DBS, cast serious doubts upon the economic viability of DBS.

DBS did not only cause complications within Britain. As DBS was to be a new kind of television and as other European nations were also drawing up DBS plans, it was seen as an opportunity to develop an advanced television transmission signal capable of carrying more information than those currently used. The problems which had arisen due to the incompatibility of the British PAL transmission standard and the French SECAM standard had led the European nations to hope that a common European standard could be developed.

The IBA had developed a high quality transmission system based on a multiplexed analogue component (MAC). This standard became the basis for a common European standard, although variants soon developed, each with its strengths and weaknesses. The British government favoured C-MAC which was capable of carrying 20.25 megabits of data per second. The French and Germans were in favour of D2-MAC which was capable of carrying only half as much information, but which used a smaller signal which could be carried on older cable systems. The delay in arriving at a common European transmission standard in effect meant that each nation's individual DBS effort was delayed. No nation wanted to take the risk that its DBS signal might either be obsolete or the odd man out of the European system.

To further compound an increasingly untenable situation, in September 1983, the Home Secretary announced that he would allow the private sector led by the IBA to compete against the BBC in DBS. Although the IBA would not get the same exclusive terms as the BBC it would be able to bid for 2 DBS franchises either on its own or as part of a consortium with others.

Although a competitive DBS was not nearly as attractive a proposition as a monopoly DBS, the BBC and IBA were able to produce a compromise plan which consisted of three channels using two satellites jointly controlled by both authorities.

Even after this restructuring and influx of private money, Unisat was still thought to be twice as expensive as existing, though non-British, alternatives and not viable without substantial government assistance. In May 1985 the Unisat project was abandoned.

The IBA, however, still believed that under different conditions a British DBS venture could survive. Their arguments, along with others from the private sector persuaded the Home Secretary to revive the DBS project. If DBS were to survive this time, however, it was felt that there could not be as many conditions on the nature of the project.

In the new project specifications the IBA was made the governing authority. They would issue the franchise and monitor its development. Although the C-MAC transmission standard was still a part of the specification, the necessity for an extremely powerful 230 watt satellite has been discounted. With satellite receiver technology advancing rapidly, it will still be possible to pick up the signal of a satellite transmitting at half this strength with a dish measuring only about 45 cm in diameter. The length of the franchise contract was extended from 10 years to 15. Perhaps the most significant change in the project specifications has been the dropping of the requirement that a British satellite system be used. The thrust of UK DBS had shifted from promoting national goals to economic viability.

The IBA's advertisement for the franchises brought forward five applicants from a range of media fields. The successful applicant,

Table 5.5: The British Satellite Broadcasting (BSB) DBS Consortium (July 1987)

BSB members	Amount contributed to first round financing £m
Alan Bond	50
Granada Group	35
Pearson	30
Virgin	25
Chargeurs SA (a subsidiary of J. Seydoux)	24
Reed International	20
Anglia TV	11.5
London Merchant Securities	10
Next	10
Investment International SA	5
Trinity International Holdings	2
	£222.5m

British Satellite Broadcasting (BSB), was announced in December 1986. BSB was chosen for its proposed television schedule which mixes public service with general entertainment, for its combination of marketing and media expertise and for the dominance of British companies within the consortium.

BSB is a consortium which was originally made up of Granada, Virgin, Amstrad, Pearson and Anglia Television. Numerous changes in the consortium's make-up have since taken place. Amstrad withdrew, while other companies have joined. In July 1987 when the first round of financing was completed, the consortium was made up of the 11 members shown in Table 5.5.

In late May, BSB awarded the £200m ($304m) contract for the provision of two DBS satellites to the American company, Hughes Aircraft. The competition between Hughes, Comsat and British Aerospace had been fierce. Comsat had hoped that the discount they were offering on two second-hand RCA satellites, left over from the failed American DBS attempt, would prove attractive. British Aerospace had been hoping to receive the contract, particularly since the initial DBS guidelines called for a British manufacturer to build the satellite.

In the buyer's market which existed in the mid-80s, Hughes was able to put together the best finance and launch date package. Hughes will deliver BSB two HS 376 satellites in orbit. By delivering the satellites in orbit, BSB will not be financially liable until the satellites are operational. The transmitting strength will be 120 watts.

As part of their package, Hughes guaranteed an early launch date. The first launch is due in the fourth quarter of 1989. It will probably be in October. The launch is scheduled to be the first commercial launch on the McDonnell Douglas Delta rocket recently brought back into service for the US Air Force. BSB's first satellite will use a current Delta rocket, the second a Delta II. BSB sees the Deltas as more reliable than the European Space Agency's Ariane rocket.

With the choice of an American company to supply the British DBS satellite, the issue of national industrial and communications policy was again raised. Two British trade unions, the Association of Cinematograph, Television and Allied Technologies (ACTT) and TASS, have lodged complaints that an overseas firm will have control over BSB policy. BSB has denied these allegations by pointing out that the Hughes loan is not convertible to shares.

BSB proposes to run four services on its three channels. These will consist of *Screen* — a film channel, *Zig-zag* — a children's channel, an entertainment channel and a news service developed and provided by Independent TV News. Given the increased competition BSB expects to face from Astra and Eutelsat II, the consortium requested that the Home Office allow it to use all five channels allocated to Britain by WARC '77, instead of just the three franchised by the IBA. This request was rejected.

By being able to be received on small TVRO's costing about £250, DBS programmes will be within reach of almost all TV viewers, whether or not they receive cable television. In many sectors the future development of cable and satellite is phrased in terms of a competition rather than the mutually beneficial relationship which exists today. Potential new media viewers are asked to choose between DBS and cable, while satellite programmers and promoters speak of a post-cable era. Most representatives of the cable industry, however, do not view DBS as competition; they continue to see it as a partner. The cable industry's argument is that since DBS will only provide three channels with a maximum of five, the potential cable subscriber will still have a strong incentive to take cable where not only will the DBS channels be available, but also all of the other existing satellite channels which the small DBS TVRO is not sensitive enough to receive. The feeling is that where cable television is already a presence, it will flourish, particularly as more satellite channels come on stream. The establishment of new cable franchises will become increasingly difficult, however, as DBS takes hold. But cable will not only have to contend with the three extra channels transmitted by BSB, but also with the 32 channels envisaged for the coming pan-European medium power satellites.

MEDIUM POWER SATELLITES — THE UNKNOWN CHALLENGE

Now that ten years of planning a commercially viable DBS system have passed, its success may depend less on what it has to offer and more upon the changes which have taken place in the marketplace during the interim. Medium power satellites have stolen much of the thunder of the high power satellites for national DBS.

The size of a TVRO required to receive the signal of a medium powered satellite, with a transmitting strength of 40 or 50 watts, has shrunk to the point where it need only be a few centimetres larger than that required for a high power satellite. The term 'DBS' is, in fact, now commonly used for both high-power and medium-power satellites.

This closing of the gap between high and medium power satellite reception highlights the political factors in the establishment of a high power satellite television system. The economic gains to be made by high power DBS have been considerably eroded by the competition of medium power satellites. Medium power satellites are considerably cheaper and cover a larger area with their signal. Yet, the medium power satellites do not belong to any one nation and therefore cannot be a source of national pride. In effect, high power satellites are domestic broadcasting systems which spill out while medium satellites are international systems which spill in.

Two organisations, one public and one private are currently planning medium power satellites which will transmit strong signals to the UK. The Société Européenne de Satellite, a commercial organisation based in Luxembourg, plans to launch Astra, a 16 channel satellite, in June 1988. This will be followed by Eutelsat II, launched by the public service oriented European Telecommunications Satellite Organisation.

Most programmers currently transmitting on low power satellites will switch to medium power in order to increase their audience. British programmers such as Sky, Super Channel, Premiere and The Children's Channel face direct competition from the programme line up proposed by BSB. As the launches of Astra and Eutelsat II are scheduled to be before the launch of the BSB satellite, programmers are likely to switch at the earliest opportunity to one of the medium power satellites in order to get a head start on their rivals.

The competition between BSB and Astra has caused sharp divisions between UK investors. Thames Television, the largest television company within the IBA network, has opted to become a 5% investor in Astra rather than BSB. Thames thus has a representative on the SES board.

It has recently been joined in the Astra venture by London Weekend TV, another ITV company, and Carlton Communications, Saatchi and Saatchi and Dixons, a large electronics and home entertainment retail chain. These new additions to Astra — all part of a failed bid for the DBS franchise in 1986 — will have two major consequences. Firstly, these companies plan to offer two entertainment channels on Astra; secondly, a situation will exist where ITV companies will face each other in direct competition — Thames and London Weekend TV on Astra and Granada and Anglia on BSB.

The implications of this for programming and also for the regulation of satellite broadcasting are far from clear. Will, say, Thames agree to sell its programmes to BSB? Will Granada agree to sell its material to the Thames Astra channels given that it will be a participant on the Super Channel service which will most likely also be available on Astra? Finally, how will the IBA regulate BSB as a public service facility when the same interests will compete with it through a commercial service on Astra?

The satellite broadcasting scene is also witnessing other divisions of interests. Divisions exist not only between high power and medium power satellites. Between the two proposed medium power satellites there exists an even sharper rivalry. This rivalry is fuelled in part by the different principles guiding the two satellites. Eutelsat II is a public service satellite, supported officially by the national telecommunications entities within the member countries of the EBU. Eutelsat believes that the terms of its charter with the EEC guarantee it a monopoly for satellite services. As a non-profit organisation, Eutelsat portrays itself as maintaining the public service tradition which has been central to broadcasting and telecommunications systems in Europe. Astra, on the other hand, claims that it is not bound by the stipulations of the Eutelsat charter. It has no qualms about being a purely commercial venture. There is currently a fierce war of words raging between the two organisations.

Given Eutelsat's view of Astra as an interloper transgressing its right to a public service monopoly, many were surprised when British Telecom, one of the national telecommunications entities pledged to support Eutelsat, leased 11 transponders on Astra in addition to the eight transponders for which it has an option on the first Eutelsat II satellite. An English uplink to Astra has also been arranged.

The strong involvement of Thames TV, London Weekend TV (LWT) and British Telecom in Astra and Eutelsat guarantees a strong British presence on medium strength satellites. Although SES is proposing that only four of its transponders be used for English language broadcasts, it is likely that the final number will be much higher. For sound economic reasons, all of the English language broadcasters may wish to unite on one satellite. Astra, because of its earlier launch, as well as the involvement of BT, LWT and Thames, is the likely candidate.

If all of the British channels are on one satellite, CNN and Anglovision, as American English language news channels would undoubtedly follow. CNN is already known to be favouring Astra. In 1987, this combination of British and American broadcasters would have given Astra 10 English language channels, well over half its 16 channel capacity. The battle between Astra and Eutelsat might then be turned into a battle between English language and other language programming.

THE REGULATORY ISSUES RAISED

Medium power satellites will certainly cause greater regulatory problems than high power satellites. Since high power satellites were designed to service individual countries, they may be regulated in a similar manner to other domestic broadcasters. Although BSB is currently regulated by the IBA, there would be little difficulty in switching responsibility for it to the Cable Authority or even to the Home Office. The transnational and non-governmental nature of medium power satellites and the

169

potential spill-over from national DBS ventures, however, raise a number of complex issues.

Cultural sovereignty

The issue of cultural sovereignty will no doubt be raised. Britain, by accident and design, will be at the centre of this issue. British Telecom's future control over 19 medium-power transponders has certainly not gone unnoticed. English language programmes already dominate the terrestrial broadcasters' programming schedules around the world. Given Britain's heavy satellite involvement and English's place as the most widely spoken language in the world, there is little reason to doubt that English language programmes will also reach the largest satellite television market. With both the U.S. and Britain producing English language programmes, the competition between the two is based on culture and quality. Britain is currently pitting Super Channel against Sky Channel to test the appeal of British over American programmes. Although Sky Channel currently dominates, Super Channel is still in its infancy.

International news flow

Britain is also hoping to present a strong challenge to the Americans with regard to international news coverage. The issues regarding international television news coverage are still sensitive and unresolved. As noted earlier, the American CNN was, for a time, not transmitted on British cable networks. CNN's restriction was ostensibly due to its Associate Member status in the European Broadcasting Union (EBU). While other news agency members of the EBU are able to use material from CNN, the terms of CNN's membership prohibited it from re-directing European news agency material back into Europe where it would compete with the domestic news service.

Given the EBU restrictions, CNN had to make arrangements with each member state of the EBU concerning its transmission on the domestic cable network. In the UK, domestic cable subscribers were not able to receive CNN because it would be competing with the domestic news services, particularly ITN's Super Channel News which began transmission at the end of 1986 and is also aimed at the cable and satellite market. CNN is, however, permitted in major hotels where it is assumed that guests are temporary residents and not part of the domestic news service's regular audience. CNN has now successfully re-negotiated its status in the UK.

CNN's battle for acceptance in Europe exemplifies the difficulties international television news broadcasters may face when trying to establish a pan-European service. CNN's difficulty may be that it is not European.

Britain hopes to be a major European voice in international news coverage. ITN has already put together a satellite news programme which is currently transmitted on Super Channel. It too, however, is running into numerous political obstacles. As with CNN, the terms of Super Channel's associate member status with the EBU poses Super Channel news with difficulties. Super Channel has to pay the EBU for access to the daily EBU news exchanges; it cannot show reports culled from these sources before 10 pm, and must negotiate a royalty with each member broadcaster. For these reasons, Super Channel news is being kept off the cable systems of several countries.

In spite of the regulatory difficulties of establishing an international news service there are many other aspirants hoping to do just that. The BBC is known to be anxious to establish a television complement to its world radio service and has already produced a pilot satellite television world service programme. Britain's other major international news service, Reuters Holding Plc, has sponsored the World News Network and would like to broadcast by satellite in several languages. Reuters have recently applied for a transponder on the French TDF-1 satellite. On top of this, another American service has entered the English language international television news fray. Anglovision, sponsored by the American network, NBC, has begun transmitting a pilot service on Eutelsat.

Foreign coverage of domestic events is always a sensitive subject. We may expect to see feathers ruffled on all sides of the Atlantic as differing national attitudes towards the 'news' come into conflict. In 1987 the British government waged a legal battle to halt publication abroad of *Spycatcher*, the memoirs of Peter Wright, a former British MI5 agent. This book was perceived to be damaging to the national interests and mention of the allegations made by Mr. Wright were prohibited from the domestic news media. Foreign news media transmitting directly to the British

171

householder will not be subject to such restrictions. It therefore remains to be seen what measures will be taken when foreign satellite news coverage conflicts with the domestic national interest.

Public service broadcasting

In Britain, cable and satellite television are not bound by the same public service strictures as are the terrestrial broadcasters. Much as in other countries, where once the limited number of available broadcast frequencies made it necessary to prescribe the manner in which they were used, the current abundance of channels is seen as making such regulations unnecessary. The current scenario has domestic broadcasters continuing to provide public service programming, while the new media will establish themselves by providing the type of programming which attracts the largest audience.

Since the government sees it as essential that Britain maintain a strong place in the new media environment, yet is unwilling to subsidise this new industry, allowing the new media the fullest freedom in the competitive market place is seen as the only possible policy. International satellite broadcasters, as opposed to national DBS broadcasters, face only self-imposed guidelines with regards to news content and limits on advertising and sponsorship.

Terrestrial broadcasters, which must also compete for the same audience, are finding their position increasingly difficult in the new environment. In view of the leniency shown towards new media programming, IBA stations as commercial broadcasters, are asking for a similar flexibility with regard to programme sponsorship. The IBA hopes that with the re-election in 1987 of the Conservative government, policies will continue towards a greater commercial freedom for the broadcaster.

The BBC's funding structure is also being scrutinised as a result of the advances in satellite and cable technology. The 1986 Peacock report on the funding of the BBC proposed that as the technology becomes available, the BBC should adopt a pay per view (PPV) funding structure.[5] The Peacock proposals were originally made with regard to the possibilities opened up by a national cable grid. With the cabling of Britain taking place much more slowly than anticipated, methods of PPV via satellite are now being considered. Although the exchange of the licence fee for a PPV funding structure is still a distant prospect, the ramifications of PPV on the BBC's ability to continue fulfilling its public service obligations are currently being debated.

PUBLIC ATTITUDE

Although the media journals are awash with stories of a media 'revolution', the British public has been slow to see and accept the changes taking place. Teletext and videotex, available since the 1970s, are only now beginning to take hold. Cable television, as already described, has had a bumpy start. Only video cassette recorders have had any significantly rapid acceptance by the public.

Rather than a reluctance to accept new technology, the quirks apparent in the British public's acceptance of new television technology may have less to do with technology and more to do with the television side of the equation.

One aspect of the British attitude towards television is a strong loyalty to the existing domestic services. Annual surveys conducted by the IBA reflect a high level of public satisfaction with British television. One recent study by the Broadcasting Research Unit identified a segment of the population which are extremely protective of the BBC as a British institution and suspicious of anything, even other terrestrial channels, which might harm it in any way.[6] Loyalty to the BBC is thus, among some viewers, equated with loyalty to Britain.

This ties in with another British attitude, annoyance at receiving a channel which they don't watch, particularly if the channel is in a foreign language. Some cable franchisees believe that this attitude has affected subscribership. Cable television marketing had been structured on the US tiering approach. Subscribers were offered groups of channels for a package price. A number of subscribers objected, however, to paying for channels in the package which they did not intend to watch. In light of this, some cable franchisees have switched to an *à la carte* approach, allowing the subscriber to pay for and receive individual channels rather than packages of channels. Satellite television may face a similar difficulty by offering the viewer more than he wants to receive.

It is perhaps with respect to these aspects of the British national character, loyalty to the existing services and annoyance with foreign programming, that the earlier cited CIT model may be faulted. The CIT model is a general model meant to predict a general trend throughout Europe. As such, it may be the case that cultural factors specific to the individual countries, were not taken into account. The British ethnocentric attitude towards television may be a strong factor in the speed of satellite television's growth in Britain. If satellite television is seen as invading the household with too many unwanted and un-British programmes, penetration may be slower than expected.

CONCLUSION

Although the public's attitude towards satellite television may be somewhat cooler than expected, the extent and impetus of Britain's involvement in satellite television ensure that the UK will be a major player in the European satellite television environment. On the hardware side, British Aerospace is Europe's major constructor of communications satellites. On the programming side, Britain, currently with eight satellite channels, is strongly influencing the flavour of satellite television.

The British DBS effort is now well on track for a 1989 launch, with hopes of establishing a new phase in British broadcasting. Its appeal as a British satellite may overcome some of the cultural qualms certain sectors of the public may have about satellite television.

The government, for the moment, appears content to monitor the changes taking place in the new media without putting forward any clear cut policy on the direction these developments should take. The policy is rather to react as changes occur.

By the mid-1990s the media environment will be completely changed, with both a new regulatory framework for domestic broadcasters and a whole array of broadcasters outside the reaches of any one country's domestic guidelines. Given the rapid changes taking place both in technology and policy it is probable that only then will all the pieces fall into place.

REFERENCES

1. Chin-Chuan Lee, *Media Imperialism Reconsidered*, Sage, Beverley Hills, 1983, p. 78.
2. cited in *The Public Service Idea in British Broadcasting — Main Principles*, Broadcasting Research Unit, London,1986, pp. 25–26.

3. Information Technology Advisory Panel (ITAP) *Cable Systems*, London, 1982, pp. 7–9.

4. Tydeman, J., *Analysis of Existing Independent Forecasts of the European Market for Cable, SMATV and Direct-to-Home Reception, 1987–1996*, Société Européenne des Satellites, Luxembourg, 1987, p. 28.

5. *Report of the Committee on Financing the BBC*, Cmnd.9824, 1986, p. 146.

6. Morrison, D., *Invisible Citizens: British Public Opinion and the Future of Broadcasting*. The Broadcasting Research Unit, 1986, pp. 32–37.

6

Satellite Broadcasting in France

Raymond Kuhn

INTRODUCTION

As broadcasting systems throughout western Europe continue to
evolve in the light of new technological, social and political
developments, France stands out as a leading example of radical
change. In the space of the past few years the French broadcasting
system has altered beyond recognition. It is not just that new radio
stations and television channels have been established in a
remarkably short space of time, though this itself would have been
significant. More importantly, the quintessential nature of the
system has been transformed. The French state has abandoned its
monopoly control of broadcasting and opened up the field to a host
of new forces of which commercial media entrepreneurs are only
the most evident. Consequently, it is within this overall context of
a national broadcasting system which is still going through a
revolutionary transition that the anticipated impact of satellite
technology must be placed.

In itself, however, the purely national context is insufficient. The
development of satellite broadcasting in France must also be
viewed against the backdrop of wider changes taking place across
western European broadcasting. While the pace of these changes
may vary from one national broadcasting system to another, the
general thrust embodies certain common characteristics. These
include the end of the domination of national broadcasting systems
by public service monopolies; the search for alternatives to
licence-fee funding; the spread of privately owned television
channels financed from commercial advertising; a growing crisis in
domestic programme production; an increase in the amount of
television available (more channels, longer schedules); the advent

of alternative sources of programming to traditional off-air television, including cable, video cassette recorders and satellite transmission; a trend towards deregulation of output; the spread of transnational broadcasting; and the incursion into this changing broadcasting environment of powerful new actors in the shape of multi-media companies operating simultaneously in and across different national systems. The broadcasting revolution in France thus forms part of a broader seismic upheaval in the field of European audiovisual media.[1]

The overall aim of this chapter is to examine French government policy on direct broadcasting by satellite (DBS) and to assess this policy in the light of developments in the French and European audiovisual systems. The chapter concentrates on DBS because other forms of satellite transmission which reach the viewer indirectly, such as those beamed down to a headend and routed via cable networks, have so far been of little import in France. The chapter is divided into three parts: first, a section on the changing French broadcasting system; second, a section on the DBS policy debate; and finally a section on French media policy in the 1980s.

THE CHANGING FRENCH BROADCASTING SYSTEM

Prior to the election of Francois Mitterrand as President of the Republic in 1981, the French broadcasting system had undergone relatively minor changes since the end of the second world war. The traditional postwar system was a highly centralised state monopoly under close and frequently direct governmental control. This meant first that television was used as a nation-building device in this diverse and fragmented society, with only minimal token concessions to local customs and regional demands. The output was overwhelmingly dominated by Paris. Secondly, the state broadcasting services, particularly television, did not have to face up to any commercial competition, even of the regulated, public service variety which breached the BBC's monopoly in Britain in 1954. While in some frontier departments viewers could receive programmes transmitted from foreign channels, in most parts of France the state television services had a captive, if not always particularly satisfied, audience. Finally, political coverage was manipulated for partisan ends, ranging from short term electoral manipulation to long-term ideological indoctrination.

177

Appointments to key decision-making posts were made more on political than professional grounds.[2]

Of course, between 1945 and 1981 important changes in French broadcasting did take place. The slow spread of television in the 1950s was followed by its gradual penetration as a mass medium in the following decade and culminated in saturation coverage of the country by the 1970s. The establishment of a second channel in 1964, the introduction of commercial advertising on the two state channels in 1968, the creation of a third channel with an allegedly regional vocation in 1973 and the attack on the state broadcasting monopoly by local pirate radio stations in the late 1970s were all evidence of a broadcasting system growing in size and complexity and to some extent seeking to satisfy new demands.

To help cope with these and other changes, major broadcasting statutes were introduced in 1964, 1972 and 1974 respectively. These pieces of legislation had important consequences for the organisation and management of the broadcasting services. For example, President Giscard's reforms in 1974 dismantled the unitary structure of the ORTF, the state broadcasting corporation set up by the Gaullists ten years previously. This reform isolated the different functions of the television system and set up separate companies to handle each aspect, including a production company (SFP), a transmission company (TDF), a radio company (Radio France) and three television companies (TF1, Antenne 2 and FR3). No subsequent reform has sought to go back to the concept of a single broadcasting corporation. Nonetheless, the substantive differences in content between the statutes introduced under the Gaullist, Pompidolian and Giscardian presidencies were far less important than their shared objective of ensuring the state's continued domination of the broadcast media. All of them, therefore, maintained support for the legal framework of the state monopoly, while in their implementation none seriously challenged the ingrained practice of partisan political control.[3]

The result of largely incremental change and lengthy bouts of consolidation during the postwar period, this system has been virtually swept away over the past five years. The ground for this paradigmatic shift was prepared by the 1982 statute on audiovisual communication, the spirit and letter of which can be regarded as both a reaction to perceived defects in the traditional system and a preparation for France's entry into the much-vaunted information technology era. The principal feature of the 1982 Socialist reorganisation of broadcasting was the abandonment of the state monopoly. At a stroke the state had finally relinquished its jealously guarded exclusive hold over broadcasting.[4]

The consequences of this dismantling of the state monopoly were not slow to make themselves felt. Towards the end of 1984 a new off-air television channel was established, the first in nearly twelve years, to compete against the existing three state networks. While initially the state was a powerful indirect shareholder in Canal plus, the fourth channel soon came to be owned by a majority of private shareholders. The financial arrangements of the new channel also marked a break with the past. Canal plus is funded not by the traditional sources of licence and advertising revenue, but rather by monthly viewer subscriptions. It is this subscription system which clearly distinguishes the newcomer from all other off-air television channels in France, including those set up in 1986. To enable it to attract subscribers France's first pay-TV channel was bound by fewer programming constraints and public service obligations than its state competitors. With its schedules overwhelmingly dominated by feature films and live sporting events, Canal plus has concentrated on popular entertainment television aimed at the mass middle-brow audience, a marketing strategy which after early teething problems has so far proved remarkably successful. Indeed, the funding of Canal plus may well become a model for other television channels in western Europe, including those transmitted by satellite.

The commercial success of Canal plus looked far from certain in the early months of 1985 when President Mitterrand unexpectedly announced his support for the creation of off-air commercial channels which would be free to the viewer at the point of reception. The necessary technical and administrative arrangements were hastily made — indecently so, many claimed, and not just amongst the parliamentary opposition — so that two private channels could be in operation before the 1986 parliamentary elections. Mitterrand seized the opportunity to open up television to new economic and social interests, just as he had legalised private local radio at the start of his presidency, grabbing whatever kudos he could from the initiative. At the same time, by filling up the airwave spectrum he hoped to limit the freedom of manoeuvre of the current right-wing government. The Socialists had thus been responsible for the introduction of both pay-TV and commercial television to France within a period of less than eighteen months.

A general entertainment channel, La Cinq and a specialist music channel, TV6, duly began transmissions in the spring of 1986. La Cinq was jointly owned by the Italian media magnate Silvio Berlusconi whose Finivest company possessed 60% of the shares and a French businessman and personal friend of President Mitterrand, Jerome Seydoux, whose Chargeurs company owned the other 40%. The main shareholders in TV6 were the advertising company Publicis and the film distribution company Gaumont.

Nonetheless, the President's political manoeuvre was only partly successful. After the elections the government of the Right led by Prime Minister Chirac introduced its own reform of broadcasting — the communication statute of September 30 1986.[5] The franchises awarded by the Socialists for channels five and six were annulled and a new regulatory body, *la Commission Nationale de la Communication et des Libertes* (CNCL) was set up with one of its early tasks being to adjudicate on a new allocation. The CNCL gave control of channels five and six to well-known political sympathisers of the Right, with only one of those favoured by the Socialists, Berlusconi, retaining a share in the fifth channel. In a politically controversial move the most important newspaper owner in France, Robert Hersant, was given a majority stake in channel five. Control of channel six, now rechristened M6, was given to a group which included the Luxembourg broadcasting organisation *la Compagnie Luxembourgeoise de Telediffusion* (CLT) in its ranks.

However, the new government of the Right envisaged a much wider sweeping reorganisation of the French broadcasting system than a mere reallocation of existing franchises. This was only an *hors d'oeuvre* to whet the appetite for the *piece de resistance* of the 1986 reform — the privatisation of the main national state channel, TF1. After a period of intense political squabbling and pressure group lobbying, 50% ownership and effective control of TF1 was given in the spring of 1987 to a group which included Francis Bouygues and Robert Maxwell among its members. The other 50% of the shares were sold off to members of the company staff (10%) and the general public (40%). TF1 thus became the first and as yet the only major state television channel in western Europe to be hived off to the private sector.[6]

Radical changes in off-air television provision during the 1980s have been accompanied by initiatives elsewhere in the audiovisual media field. One such development is cable.[7] In 1982 the Socialist government embarked on an ambitious cable plan the objective of which was to provide business and domestic users in the long term with a wide range of interactive services. More immediately cable would also give its mainly urban subscribers a multi-channel television facility. The original aim of the cable plan was to cable one and a half million households by the end of 1987 and six million households by the mid-1990s. In fact, however, because of high capital investment costs and politico-administrative conflicts the rate of cabling has proceeded much more slowly than initially projected. By the end of the decade cable television will be a reality for only a very small percentage of the French population. As a result, for the foreseeable future the vast majority of French viewers will still be dependent on programmes from television channels using the old technology of land-based transmitters — and, of course, on the output of the new satellite channels.

The result of these various additions and changes to the French television system in the 1980s has been the growth of a large privately owned commercial share of the television programming market and the concomitant decline in both relative and absolute terms of the state sector. In the summer of 1984 there were only three television channels which the majority of French viewers could receive: TF1, Antenne 2 and FR3. All three were state owned. By the summer of 1987 six channels were widely available, of which only two remained in state hands, Antenne 2 and FR3. The other four — TF1, Canal plus, la Cinq and M6 were privately owned. With the possible exception of Italy in the 1970s, the structure of no major western European broadcasting system has changed so dramatically in such a short space of time.

THE DBS POLICY DEBATE

DBS is the other principal element in this changing audiovisual landscape. French viewers will soon have the opportunity to receive programmes from four channels transmitted from the French direct broadcasting satellite, TDF1, which is due to be launched in the spring of 1988. A back-up satellite, TDF2, is scheduled to become operational soon after. Unlike the lower powered communication satellites such as ECS1 whose television output can usually be received only after retransmission via cable networks, the programmes transmitted from TDF1 will be *directly* available to viewers equipped with the special reception dish. Because of the size of the satellite footprint, viewers in neighbouring countries will also be able to tune in. Hence TDF1 marks not just an innovation in the French broadcasting system, but also a new factor in the European audiovisual environment.

As one of the major industrial and broadcasting countries in Western Europe, France has been in the forefront of those states developing audiovisual satellite technology. After all, Fifth Republic France is renowned for its commitment to high technology prestige projects, ranging across nuclear power, telecommunications, railways and aerospace. Yet as with the ambitious cable plan, the French government's DBS project has had various problems to overcome. Over the past ten years the debate concerning the development of DBS has been riddled with disagreement and dissension. Politicians, often of the same party, have taken opposing stances, government ministers have fallen out with each other, the administration has been divided, commercial media companies have competed for governmental favours and France has frequently found itself in conflict with fellow EEC member states, most notably Luxembourg, over satellite policy. At the same time techno-industrial goals have often been at odds with cultural objectives. As a result, the strategy set out in the late 1970s has been questioned and re-examined at various stages since in the light of technological advances, developments in off-air television and changes in the composition of the governing majority. In short, France has not pursued the DBS option single-mindedly or without controversy.

Rather than adopt a purely chronological approach which would inevitably replough the same substantive ground, this section will examine analytically certain key aspects of the DBS debate in France over the past decade. It will be particularly concerned with the origins of the DBS commitment, the politico-administrative conflict on the utility of DBS, the France-Luxembourg relationship and the allocation of the channel franchises on TDF1.

The launching of TDF1 will mark the culmination of a policy whose origins can be traced back to the presidency of Mitterrand's immediate predecessor, Valery Giscard d'Estaing. It was during the late 1970s that the Giscardian government agreed to combine forces with Helmut Schmidt's West German government to construct two direct broadcasting satellites, one for each country. On both sides the underlying rationale was principally techno-industrial.[8] Both governments were acutely aware of the need to invest in new audiovisual and communications technology if potentially lucrative world markets were not to be left to competing industrialised nations, particularly the United States and Japan. While the allocation of the future available channels was naturally the object of much speculation, it was evident that the French government was open-minded on what it clearly regarded as a secondary question. The overriding objective was not to miss out on the production and development of the new hardware. Hence the decision was taken to pursue the DBS option before any policy was formulated as to how DBS would be used and for whose benefit. Industrial considerations clearly predominated over broadcasting or cultural ones.

However, even the industrial argument in favour of the development of DBS did not convince everyone. One running conflict centred on the question of whether DBS was necessary if France was adopting cable at the same time. It was not apparent to everyone in government and in the upper echelons of the administration that the simultaneous pursuit of two high technology objectives in audiovisual communications was essential or even desirable. Commitment to the DBS option was most clearly and consistently manifested by the state transmission company, TDF, which has the responsibility of maintaining the terrestrial transmission networks and supervising frequency usage. For TDF DBS was a prestige project which allowed the company to present itself in the vanguard of technical progress. Within the ranks of the French government, TDF found support in the Ministry of Industry and the Ministry of Communications.

In contrast, within the Ministry of Posts and Telecommunications, and in particular inside its telecommunications wing — *la Direction Generale des Telecommunications* (DGT) — voices were raised against the DBS option, notably in the mid-1980s by the DGT's director general, Jacques Dondoux. The DGT had been responsible for the modernisation of the French telephone network in the 1970s and for the formulation of the cable plan in the early 1980s. DBS posed a challenge to its desire to control technological developments in the communications field and so was fiercely resisted. Arguments from this opposition camp emphasised the huge research and development costs of DBS, its technological obsolescence and its unproven record. It was stressed that given the Socialist government's pledge to wire up France to fibre optic cable, the state should concentrate on the lower powered and much cheaper telecommunications satellites such as Telecom 1 which could beam programmes down to headend stations which in turn would transmit them through the cable networks. State investment could thus be concentrated on one coherent technological objective, while the viewer would be spared the additional expense of having to purchase the special parabolic dish for satellite programme reception. Another advantage for the state was that this option would allow the relevant regulatory authority within France to control the viewers' programme intake by monitoring output at the cable headend. Undesirable programming could be repelled at the French borders.

The DGT/TDF rivalry looked set to continue even beyond the launching of TDF1. As recently as the summer of 1987 the PTT minister, Gerard Longuet, joined other government ministers in a critique of the TDF1 project and proposed that one of the Telecom satellites be exploited as from 1988 to transmit ten television channels.[9] Because of its capacity to cross subsidise its operations, the DGT could lease the channels to operators at significantly more attractive rates than those applicable to TDF1. The disadvantage of this project was that the cost of the reception dish to the viewer would be much higher since the satellite signal was not so powerful. Whatever the results of this particular proposal, it was clear that the DGT stood ready to profit from any last minute hitch in the TDF1 project and intended to do everything it could to ensure that TDF did not enjoy a monopoly in satellite broadcasting in the future.

This conflict within the upper echelons of the French administration was largely fought out between technocrats, or even technicians. Ministers also became involved as representatives of their respective departments, but as one might expect the issue never became one of intense public controversy. Compared to the furore surrounding the establishment of channels five and six or the privatisation of TF1, the debate about new audiovisual technology hardware was largely conducted by insider specialists. The highly technical nature of the controversy helped keep it out of the public political domain for much of the time. But this was not the only factor. The issue also cut across party political divisions, with both Right and Left backing in a general fashion France's commitment to high technology research and development.

The decision to maintain an active interest in DBS may have been taken by persons with an archetypal technocratic training. However, it was not made with reference to technocratic criteria alone. It was essentially a political decision taken at the top levels of the Fifth Republic's policy making hierarchy. In 1984 it was the Prime Minister, Laurent Fabius, and two years later his successor,

Jacques Chirac, who reaffirmed the country's continued commitment to DBS. The need for France to maintain a presence in a potentially lucrative export market, particularly to third world countries, certainly played a part in the government's thinking. So too did the recognition that whatever policy France adopted, the state would be unable to stop output from other direct broadcasting satellites from overspilling into French territory. In any case it would have been unthinkable for France to have dropped out of the DBS race at a time when after vacillation and hesitation in some quarters all other major industrial states showed every intention of competing. Better to join in the DBS game as a participant, the French government seems to have concluded, than leave a potential vacuum to be filled by others.

The pursuit of the DBS option has been dogged by technical problems, including the failure of the Ariane launcher to get off the ground. However, it is the political problems which have remained at the forefront of the satellite debate. One of the thorniest of these has been the uneasy relationship between France and its media neighbour Luxembourg. The position adopted by Luxembourg regarding satellite broadcasting has been an important element in French thinking throughout the 1980s. For much of the decade the French feared that Luxembourg might launch its own satellite and so compromise the commercial viability of TDF1. In addition, the Socialist government in France regarded a Luxembourg satellite as an undesirable alien which would bombard France with US programming — 'Coca Cola culture' — from the skies. To prevent such a situation arising the Socialist government naturally conducted bi-lateral negotiations with its Luxembourg counterpart. In addition, through shareholdings which it indirectly controls in the Luxembourg broadcasting organisation (the CLT), the French government sought to apply pressure on the Luxembourg authorities to participate in TDF1 rather than pursue their own separate project. At one stage Luxembourg was to be wooed on to the French satellite through the CLT being allocated two of the four available channels.

While agreement on this policy was being reached in the mid-1980s, the Luxembourg government was simultaneously pulling the rug out from under the French (and incidentally from under the CLT) by setting up a new body, *la Societe luxembourgeoise de satellites* (SLS), to investigate the possibility of launching its own DBS satellite backed by US private interests. Relations between the French Socialist government and the

Christian Social government in Luxembourg reached a low ebb, despite the latter's assurance that there would be no French-language commercial television programming on the future Luxembourg satellite. This conflict dominated relations between the two governments during the summer of 1984. It was temporarily resolved later that year following a change of government in Luxembourg when a joint agreement was approved which envisaged the cooperative use by both countries of the French DBS. The Luxembourg authorities agreed not to transmit for a period of five years by means of another satellite any programmes financed by advertising (and so available to the viewer without any additional payment) which might compete with those of the CLT on TDF1. In return, the two channels of the Luxembourg broadcasting organisation on the French DBS would enjoy a monopoly position with regard to finance from commercial advertising.

However, the decision in 1985 by President Mitterrand to establish two new off-air commercial channels at the start of 1986 threw everything back into the melting pot. Relations between the two governments again deteriorated and reached a new low ebb when the CLT did not win one of the new franchise allocations, losing out to the Berlusconi-Seydoux consortium for control of channel five. The CLT's association with Rupert Murdoch seemed to have influenced the Socialist government's decision. Around the same time the Luxembourg government was giving the go-ahead to the launching of a medium powered satellite, ASTRA, which would transmit on sixteen channels directly to homes, to cable systems and to SMATV installations. While the reallocation of the franchises of channels five and six at the start of 1987 gave control of channel six to a group which included the CLT, France had failed in its attempt to keep Luxembourg out of the satellite broadcasting race.

As the launch date for TDF1 approached, attention increasingly turned to the question of the allocation of the channel franchises on the French DBS. This was scarcely a new area of speculation. Ever since plans were first unveiled backing the construction and launching of a domestic direct broadcast satellite, the issue of which companies would be awarded control of the available channels has remained an item high on the audiovisual political agenda. Over the past ten years it would appear that every possible option has been canvassed at some time or other. For example, it was initially argued by some that the satellite should transmit the output of the then two national state channels, TF1 and Antenne 2. Not only would this have ensured 100% coverage of the whole territory, but it would also have done away with the need to maintain a whole system of expensive land transmitters. Others argued for something different on TDF1 to maximise its audience impact. Several French and foreign media companies expressed an interest in participating in the use of the new technology.

At the end of 1984 the Socialist government decided that something new was indeed required to attract the public to DBS. It proposed that one of the channels should transmit for the most part original programmes, derived largely from co-productions with other European broadcasting services. Responsibility for running the programming side of this channel was given to Pierre Desgraupes, a former director general of Antenne 2. The other spare channel would be used to re-transmit the best programmes from the French and other Francophone television services. This second format is already in operation in western Europe in the form of the TV5 channel (not to be confused with the off-air channel five, la Cinq), which is at present transmitted from a low powered communications satellite and therefore available mainly on cable systems. Using this transmission method TV5 had an estimated European television household penetration of 2.5 million at the end of 1985. The Socialists later agreed to an English language channel to be run by the Robert Maxwell group and the transmission by TDF1 of the Berlusconi-Seydoux channel five.

However, when the Right returned to power in the spring of 1986 all previous decisions were declared null and void. At the time of writing (July 1987) the definitive allocation of the four available channels on TDF1 had not yet been finalised. However, some decisions had been taken and other options discarded. One channel has been allocated by the government to a public consortium comprising Antenne 2, FR3 and *la Societe d'edition de programmes de television* (SEPT). Funded out of licence revenue this channel will transmit cultural programmes during peak viewing hours. It is intended that in time it will become a European cultural channel. The German government was quick to back the French decision at the Franco-German summit in May 1987 and expressed its desire that the German public service channels ZDF and ARD should contribute to the programme output. Outside of peak hours programming would be in the hands of a different company, with Canal plus and Hachette possible candidates for the franchise.

187

The allocation of franchises on the other three TDF1 channels is the responsibility of the CNCL. Nine groups put forward applications for consideration: the four French commercial channels (TF1, Canal plus, la Cinq and M6); Visnews (Great Britain); Bravo TV (USA); Olympia (Great Britain); the German Post Office; and a French music company, Mood Music. The off-air channels five and six proposed to use the satellite to improve their national transmission coverage which has been constrained since their establishment by the scarcity of the frequency spectrum. These channels would be financed from advertising. On the other hand, TF1 and Canal plus proposed different formulae from their land based transmissions, with the former advocating a cultural channel and the latter a family viewing channel. Both of these channels would be financed from viewer subscription. The applications of Visnews, Bravo TV and Olympia were rejected by the CNCL during its deliberations in July 1987. However, the commission refused to make any final selection until the government had resolved certain outstanding problems.

The first of these concerned the financing of the DBS project. The French government was willing to cover the total costs of TDF1 (research, development, construction and launch), provided that the private sector operators paid the costs of the necessary back-up satellite TDF2. However, the prospective operators all considered the financial outlay involved too great and still hoped to persuade the government to increase state investment in the project. This demand was in turn being resisted by government ministers who argued for the application of the laws of the market in the operationalisation of the French DBS.

The second problem concerned the regulations governing the screening of encrypted or scrambled programmes. Just as it had done before the establishment of the off-air Canal plus, the French film industry was seeking guarantees regarding the screening of feature films on encrypted channels. While the potential applicants for DBS channels in turn wanted preferential arrangements so as to attract subscribers.

FRENCH MEDIA POLICY IN THE 1980s

There remained, therefore, technical, political and administrative problems to overcome before the French DBS could become operational. However, important as these problems were, none looked to be insuperable or sufficient to cause the abandonment of the project. Moreover, there were powerful reasons for France to go ahead with DBS since to give up so late in the day might have serious consequences. Politically, it would harm the Franco-German relationship which has been so carefully nurtured by both sides since the 1950s. Commercially, it would be a major blow to European industrialists keen to use the D2 Mac Pac transmission standard in their competition with the Japanese. Technologically, it would retard development in Europe towards the new high definition television.

It seemed reasonable to expect in mid-1987 that an agreement would be satisfactorily concluded between the government, TDF, the prospective channel operators and other interested parties. It could also be assumed that the Ariane launcher would successfully put the French DBS into its geostationary orbit and that the satellite(s) would function correctly. If so, television programmes would begin to be received from TDF1 some time in 1988.

The impact of the French DBS is as yet, of course, an unknown quantity. At one level the arrival of TDF1 can be regarded as another element of change in the French broadcasting system of the 1980s. Indeed in this context the impact of TDF1 cannot be divorced from the consequences of the radical changes which have recently taken place in off-air television. DBS may well extend the domestic geographical coverage of the two new off-air channels, la Cinq and M6, to the whole of metropolitan France. It will give *all* French viewers new television channels and additional programmes, albeit involving extra financial outlay for the purchase of the dish and subscription to certain channels. The number of programme sources will be increased to an unprecedented level and more viewers will spend more time watching more television. French advertisers will benefit from yet more audiovisual media to be exploited and domestic programme makers will have more outlets for their product. The recent evolution in French television away from state control and towards private commercial ownership will be confirmed. In short, the impact of TDF1 for the French television system will be considerable and, in the eyes of many, beneficial.

To confine oneself solely to the French context, however, would be seriously to underestimate the novel role of DBS. DBS is not just a delivery system for the more efficient transmission of programmes to the same national audience as was served by land-based transmitters. Nor is TDF1 merely a source of additional television programmes for French viewers. Rather, the principal innovatory aspect of DBS is its capacity to penetrate previously more or less secure national boundaries. Programmes transmitted from TDF1 will be directly available to audiences in other European countries to an extent undreamt of previously, while French viewers will be able to receive programmes from other European satellites. Thus DBS is a supremely transnational medium of audiovisual communication.

Because of this transnational character new actors and issues have become prominent in the audiovisual media debate in western Europe. Traditionally the dominant actors have been national governments, national broadcasting organisations and to a lesser extent national regulatory authorities. As was the case in France, the national broadcasters have frequently benefited from a state broadcasting monopoly and usually have paid at least lip service to the ethos of public service. The finite nature of the airwave spectrum resulted in broadcasting being treated as a scarce national resource which had to be subject to regulation in the public interest. Consequently where privately owned television was allowed in western Europe, as in the UK, it was nonetheless subject to similar regulations governing output as its publicly owned rival.

These national actors are still extremely influential. However, new transnational and supranational actors are challenging their previous hegemonic position in policy making for western European broadcasting. For example, the European Commission has sought to carve out for itself a regulatory role at the supranational level. A recent green paper of the Commission on the subject of cable and satellite broadcasting proposed the creation of a 'common market for broadcasting' within the EEC, an objective which necessitated the harmonisation of existing national legislation. In effect, the Commission was seeking support for the imposition of minimal regulations across the member countries on selected issues relating to transnational broadcasting, concentrating initially on those regulations governing television advertising. The EEC has also become involved in helping to promote co-productions between broadcasting organisations of member countries. More generally, issues which were previously dealt with inside relatively closed national systems have clearly taken on a transnational dimension. These include the regulation of broadcast output, provisions governing the amount and type of television advertising, measures to stimulate programme production and decisions on transmission standards.[10]

Another new actor on the western European media scene is the multimedia transnational commercial enterprise. A variety of multimedia companies usually associated with an individual entrepreneur have been securing a stake in off-air, cable and satellite television transmitting both in and across different national systems including France. One of these is Berlusconi who owns much of private off-air television in Italy, has a stake in the commercial fifth channel in France and is seeking to infiltrate into West German, Greek and Spanish television. Another is Robert Maxwell who has important newspaper holdings in the UK, a stake in British cable television and is now a partner in the group which owns TF1. A third is Rupert Murdoch, who owns a growing section of the British press and also controls Sky Channel which transmits general entertainment programmes to continental European audiences principally in Scandinavia, Belgium and the Netherlands via the ECS1 satellite and local cable networks. All of these and other similar entrepreneurs are buying into the new European broadcasting markets created by developments in audiovisual technology.

The interdependence and frequent tension between the national and international dimensions is crucial to an understanding of French government policy on broadcasting and the new audiovisual media during the 1980s. For example, when it came to power in 1981 the Socialist government was anxious to preserve and foster as much as possible the role of French broadcasting as a disseminator of the national culture. It was, therefore, strongly opposed to a free-for-all in the development of new media in Europe which in its eyes would lead to European broadcast outlets being used as channels for the transmission of low quality output, notably from the US. The opposition to France becoming a dumping ground for programmes reflecting an American 'Coca Cola culture' was strongly articulated in ministerial pronouncements in the early period of the Socialist administration, while the fear of a Luxembourg satellite acting as a clearing house for foreign and especially American television programming was omnipresent in the bilateral government negotiations during the first half of the decade.

Certainly even in the early period of their administration the Socialists accepted the need for change in the French broadcasting system and recognised that France could no longer be isolated from developments in the wider audiovisual world. The abandonment in the 1982 statute of the state monopoly in broadcasting was a recognition of how much had altered. The state monopoly was becoming increasingly difficult to justify in technical terms, since the old argument of airwave scarcity was much less convincing in the new world of satellite transmission and fibre optic cable. The state monopoly was also becoming harder to defend politically, with new social, economic and political actors clamouring for a stake in broadcasting. While in opposition the Socialists themselves had taken up arms against the state monopoly and supported the cause of local pirate radios during the latter part of the Giscardian presidency. Understandably, the legalisation of these radio stations had been an early priority for the Socialist government in 1981. Television was the next obvious area to be opened up.

The Socialists, however, wanted to avoid the anarchy of the Italian experience in television which had seen in its initial stages the proliferation of hundreds of low quality private channels and then the concentration of these stations in a *de facto* private network under the control of Berlusconi.[11] Their stated objective was one of controlled liberalisation of the French television system. This objective was an integral element in the Socialists' policy on audiovisual media which pursued several goals simultaneously.

One was to introduce more pluralistic control of television and wider access to the medium. In addition to the previously all-powerful state, commercial companies and sub-state actors such as local authorities would be given an important role to play in the management of television programming. A second objective was to stimulate the domestic programme production industry. Traditionally the French state has adopted a variety of measures in this regard including state subsidy and quotas on French programming. Defended on the grounds of protecting French culture and aiding employment, such provisions were enhanced by the Socialists. At the same time, in recognition of the changing international dimension the Socialist government called for a European solution to the problems posed by DBS, including the creation of a European support fund for the television programme production industry and greater emphasis on coproductions at the European level. A third policy goal was to ensure that France remained in the vanguard of the industrialised nations developing new communications technology. The cable plan in particular was regarded as of vital importance for the long term regeneration of the economy.

The Socialists hoped that these objectives could be coherently and simultaneously pursued. However, a policy which had to reconcile the cross-cutting tensions between national and international pressures on the one hand and techno-industrial demands and broadcasting/cultural considerations on the other was always going to be a difficult balancing act to sustain. It soon proved to be impossible. What began as an apparently coordinated plan for the development of audiovisual media hard- and soft-ware came to resemble a series of uncoordinated decisions which in the eyes of some were mutually contradictory. To be fair to French policy makers the information on which many decisions were being made was imperfect. Should the state back cable in preference to DBS, as seemed to be the case in the early years of the Socialist administration? Were cable and satellite broadcasting complementary or rival media? How quickly could the cable plan be implemented given the Socialist government's desire to involve local authorities and other sub-state actors as part of its commitment to the *grande reforme* of the new régime - the decentralisation of power to the regions and localities? Could undesirable foreign programming from non-French satellites be kept away from French viewers? There were no clear and easy answers to these questions and hence much of policy making was reduced to an industrial and cultural gamble.

The slower than projected implementation of the cable plan and the impossibility of controlling the activity of other sovereign states such as Luxembourg were severely hampering the successful realisation of Socialist audiovisual media policy. So too was the poor condition of the domestic programme production industry which despite government assistance was clearly incapable of responding to the challenge posed by demands for more television programmes to fill new schedules. The initiative taken by President Mitterrand to set up new off-air commercial channels subject to considerably less stringent regulations on output than their state competitors tolled the death knell of the already faltering policy. In what way a predominantly foreign owned channel five broadcasting a large proportion of non-French produced output differed from the previously much reviled and feared Luxembourg satellite channels of the future was difficult to explain. For all the early grandiose talk of a communications plan, by the end of their five years in government Socialist media policy had been reduced to short term decisions based on electoral expediency.

In many respects the development of DBS and of other audiovisual media outlets has proved less of an ideological problem for the Right. More wholly committed to the liberal doctrine of the

193

free transfer of information across national boundaries, more sympathetic to the desires of commercial media entrepreneurs and more trusting in the market mechanism to balance out audience demand and programme supply, the Right have gone further than the Socialists would have done in liberalising the French audiovisual media. With the zeal of a new convert, the Right have switched from being staunch defenders of the faith in the state broadcasting monopoly to being ardent believers in the new credos of privatisation and deregulation.[12]

The new audiovisual landscape in France, including DBS, will be welcomed by many advertisers, media companies, programme makers and viewers for the clear opportunities it provides in terms of expanding markets and range of channels. Without doubt the possibilities of pan-European television are potentially very exciting, particularly to a nation more used to a government controlled state monopoly system. For example, political information is one area which should benefit from the changes introduced in the 1980s. However, there will be drawbacks as well. On the programming side the spread of more commercially oriented, less regulated channels may well lead to yet another diminution in the amount of quality drama and fiction produced for French television. There will probably be greater recourse to foreign and especially US imports as channels seek to minimise costs in what seems certain to become a competitive struggle for the viewer's subscription and the advertiser's account. One of the new off air/satellite channels may well go out of business. In addition, many viewers will find the sheer quantity of television advertising repugnant, what little is left of the state television sector may well be deliberately starved of funds and in the increasingly cost conscious environment there will be little incentive for programme makers to take risks or produce programmes for less well off minority groups. Greeted enthusiastically by some, the radical changes in French television will be viewed cautiously by others. But is that not always the case with revolutions?

REFERENCES

1. On changes in European broadcasting systems and the new media in Europe, see R. Kuhn (ed), *Broadcasting and Politics in Western Europe*, Frank Cass, London, 1985; D. McQuail and K. Siune (eds), *New Media Politics*, Sage, London, 1986; and J. Tydeman and E.J. Kelm, *New Media in Europe*, McGraw Hill, London, 1986.

2. On the political history of French broadcasting up to the 1974 reform, see R. Thomas, *Broadcasting and Democracy in France*, Crosby Lockwood Staples, London, 1976.

3. On the 1974 reform, see R. Kuhn, 'The Presidency and the Media, 1974-82', in V. Wright (ed), *Continuity and Change in France*, Allen and Unwin, London, 1984.

4. On the 1982 reform, see R. Kuhn, 'France: the end of the government monopoly', in R. Kuhn (ed), *The Politics of Broadcasting*, Croom Helm, London, 1985.

5. For an analysis of the 1986 statute, see B. Delcros and B. Vodan, *La Liberte de Communication*, Documentation Francaise, Paris, 1987.

6. *Le Monde*, April 7 1987.

7. On cable in France, see the chapter by C.-J. Bertrand in R. Negrine, *Cable Television and the future of broadcasting*, Croom Helm, London, 1985.

8. See the chapter by Peter Humphreys in this volume.

9. *Le Monde*, July 14 1987.

10. See the chapter by Rosemary Hughes in this volume.

11. On recent changes in the Italian broadcasting system, see D. Sassoon, 'Political and market forces in Italian broadcasting', in R. Kuhn (ed), *Broadcasting and Politics in Western Europe*, Frank Cass, London, 1985.

12. On the audiovisual policy of the Right in the realm of satellite broadcasting, see J. Freches, *La guerre des images*, Denoel, Paris, 1986.

7

Satellite Broadcasting In Australia

Ian Reinecke

The capacity of satellites to broadcast television and radio signals
was closely monitored by engineers of the Australian Postmasters
General Department through technical studies that began in 1961.
It took two more decades for Australia to put a communications
satellite service into operation. Yet Australia possessed a
geography and demography that appeared ready-made for satellite
services. It is a continent roughly the size of North America, with
major population centres separated from each other by large tracts
of sparsely-populated land. The territory that must be traversed to
pass from one large city to another varies from alpine to desert.
Most of those large cities are strung along the eastern and southern
coasts of the continent where the biggest are conurbations of more
than four million people and the smallest a million.

Outside the major cities, large country towns dwindle to one-pub
outposts the further one travels inland from the coastline. In the
central desert areas, remote hamlets, stations and Aboriginal
reserves are separated by hundreds of kilometres of arid,
uninhabited lands connected by roads made impassable by heavy
rain for part of the year. Communications in large areas of the
inland are confined to high frequency radio links; there are
telephones connected to the public switched network and television
sets play recorded video cassettes, for there are no live
programmes. The heartland of Australia is largely inhabited by
communications outcasts.

Canada, a country with similar living standards, a comparable
population size and a geography in which its major cities were
separated by equally long, if much colder, distances, launched its
first satellite in 1962. Within a decade it was using a satellite for
domestic communications and a few years later was experimenting

with the high-powered, high frequency satellites that have become the world standard.[1] To Australia's north, Indonesia — a country whose 100 million inhabitants populate a string of major and minor islands connected by primitive communications — had a communications satellite in the 1970's.

To many external observers, the Australian reluctance to embrace satellites as a solution to the problems of communication imposed by distance and population distribution, was inexplicable. What those observers often neglected to note was the existence on the ground of a terrestrial communications network that had evolved over the course of more than a century. The network had been built at public expense with great resourcefulness since the last quarter of the nineteenth century. In the process some landmarks in the history of communications were established. The earliest was a telegraph line pushed north from Adelaide in South Australia, and south from Darwin on the northern coast, to meet in the middle. The line crossed thousands of kilometres of land that had never previously been subjected to white settlement; its poles were sometimes live trees with the branches lopped off, but it worked. By connecting to a British cable snaking down the Indonesian peninsula and across the Timor Sea, the colonial seats of administration in the south of the continent were able to communicate directly with London.[2]

By the 1960s, when the possibilities of satellite communications were just being realised, the Postmaster General's Department was installing the world's longest continuous microwave link, across the Nullabor Plain between Adelaide and Perth. Coaxial cable links between the two largest cities, Sydney and Melbourne, relayed live television broadcasts. As investment in the terrestrial public switched network accumulated, the arguments in favour of adding satellites to the mix of transmission paths were never decisive. Each report from the Postmaster General's Department that examined the viability of satellite communications during this period of expansion concluded that the economic advantage in Australia still resided with terrestrial technology. Despite the best efforts of an influential group of engineers and managers to persuade the organisation to the contrary, pressures for a satellite were resisted. After its metamorphosis from government department to statutory authority in 1975, Telecom Australia continued to maintain that a satellite system was not economically justified.

Almost a decade later, the Australian domestic satellite operator Aussat has two satellites in orbit, and a third awaiting launch. Its plans for a second generation of larger and more powerful satellites that would remain operational until the end of the century have been announced. Technologically, the first generation Australian satellite system, with some minor exceptions, is unremarkable. It employs technology developed, designed and manufactured in North America for television broadcasting and telecommunications. Minuscule technology transfer has occurred, despite the existence of government policies intended to oblige overseas suppliers of advanced electronic systems to involve indigenous companies in manufacturing. In telecommunications, the existence of a satellite system has not fulfilled the promise of telephone services in remote areas, although it has accelerated pressure for deregulating the national carrier. And in television broadcasting, a tortuous policy-making process has still not succeeded in delivering the programmes rural dwellers were told would end their isolation.[3]

Yet the introduction of a satellite system has had extraordinary effects on Australian communications. The satellite's role as an instrument for achieving deregulation of telecommunications stands on the threshold of realisation. Its existence has provoked an unprecedented upheaval in the ownership of the Australian media. And not just people living in remote areas, but those who populate the coastal and inner hinterland rural areas, are being proffered USA-style networked television as evidence of their entry into a new communications age. As an exercise in the national contemplation of alternatives and the construction of policies that meet the national communications needs of greatest priority, the process that led to Aussat is an object lesson in failure. It is a process by which technological infatuation with space satellites inexorably overwhelmed rational decision-making. Once infatuated, the politicians and public servants who determined the course of the satellite project permitted themselves no doubts or fears.[4]

It was the possibility of satellite television broadcasting, especially to remote rural areas, that became the most popular, and populist, political reason for establishing a domestic Australian satellite system. In August 1977 media owner Kerry Packer approached the then conservative Prime Minister, Malcolm Fraser, with a proposal that a domestic satellite be launched. The sympathetic ear lent by Fraser encouraged Packer to commission Canadian satellite consultant Donald Bond to prepare a report arguing the case for a domestic communications satellite. (Bond Report[5]).

On the basis of Bond's report, the government swiftly established an inquiry to examine the feasibility of the kind of system that Packer urged it to adopt. The task force was headed by the General Manager of the Overseas Telecommunications Commission (Australia)(OTC), Harold White, and was composed of senior public servants who predominantly favoured the satellite proposal. White himself had been a thorn in the side of Telecom two years earlier when his efforts in marshalling the support of his large customers ensured that OTC remained a separate organisation when Telecom was established. Telecom had no reason to suppose that its position that the satellite was too expensive to be justified would evoke empathy from White.

The report that White's task force prepared for the government as expected endorsed the idea of a national communications satellite, but it differed in significant respects from the Packer-sponsored Bond report. Where Bond had urged that a relatively low-powered satellite operating in C Band (4/6 GHz) should be adopted, the White task force proposed a more ambitious satellite, with a mix of high-powered transponders operating in at least three different frequency bands. They were C Band, mainly for television programme distribution by networks, Ku Band (11/14 GHz) for direct broadcast television and inter-city telecommunications services, and the band reserved for military communications, (7/8 GHz).

An Inter-Departmental Committee (IDC), composed of senior public servants from key areas of the federal bureaucracy was established by Fraser's Post and Telecommunications Minister, Tony Staley. Its function was to adjudicate between the different models for a satellite system proposed by Bond and White. It decided instead to present a third prototype. The IDC returned to the original simplicity of Bond's single-frequency band proposal, but selected Ku instead of C band. It became convinced that the higher band would allow almost everything that C band could do, but with some additional benefits. The higher the power in the satellite transponders, the smaller the diameter of the receiving dish on the ground. The smaller the dish, the lower the cost of the earth stations and the more quickly and widely was satellite communications likely to be adopted, the committee reasoned.

On a purely technical basis, there were disadvantages in selecting the higher frequency band. Heavy rainfall would

199

frequently attenuate the signal in northern tropical areas of the continent, resulting first in loss of sound from broadcast television programmes, and then loss of the picture entirely. Unfavourable weather conditions during digital transmission could corrupt streams of data, rendering them unintelligible. From a regulatory perspective, there was however one decisive advantage in choosing the higher frequency band over its rival, C band. Telecom's terrestrial microwave system operated in C Band, giving it prior claim for the responsibility to integrate the flow of satellite and ground-based transmissions. Telecom could advance no such claim in the case of Ku band.

The satellite system described in the IDC report became the blueprint for the request for tender documents that followed the federal government's decision to proceed with the project. When those documents were publicly released, it became obvious that the proposed system was modelled on the Hughes Aircraft series of Ku band satellites supplied to, among others, the Canadian government and Satellite Business Systems in the USA. Those satellites became the prototype for the three spacecraft Hughes designed to win the contract for the Aussat system. They each had a mix of high-powered 30 watt transponders and less powerful transponders of 12 watts. Clearly, the most commercially valuable of the two types were those operating at higher power because they could sustain a range of hitherto unexplored services such as direct broadcast television. The first satellite was launched in August 1985, and the second in November 1985. The third, referred to as "K3" is still awaiting launch in mid-1987, on the European Ariane space launch system.[6]

Before Aussat was launched, the Australian Broadcasting Corporation (ABC), the national, government-owned radio and television programme maker and distributor, already delivered some of its programmes by satellite. By agreement with Telecom Australia, the ABC provided programmes to people in remote rural areas. Many of these people were among the group of remote dwellers whose lack of television had been advanced as a major political justification for a domestic satellite. The ABC relayed its programmes to these viewers through uplinks operated by the Overseas Telecommunications Commission (OTC) to Intelsat satellites located above the Indian and Pacific Oceans. Telecom erected a string of large earth stations in remote areas from which the signal was then distributed terrestrially.

As these broadcasts were transmitted in C band there were none of the attenuation problems that were later to become characteristic of reception of signals broadcast in Ku band from Aussat. The greatest drawback of this system was however that broadcasts were not live; they consisted of cassettes of programmes that had already been broadcast in the major cities. A limitation felt more keenly by the commercial television channel proprietors was that the Intelsat/Telecom/OTC/ABC co-operative venture excluded their programmes, and allowed the publicly-owned broadcaster entry to markets denied to its private-sector competitors.

The government's decision that the ABC should take up the largest single number of transponders on Aussat spelled the end of the Intelsat arrangement. The ABC was obliged to withdraw from the system using Intelsat because of its technical incompatibility, on two levels. The receiving stations set up to catch signals from Intelsat's satellites were designed to accommodate C band traffic, not the higher Ku band used by Aussat. And its Intelsat system used the Australian standard PAL television signal format which was incompatible with the one eventually adopted by Aussat, called B-MAC. A third factor that differentiated Aussat from Intelsat was the standard earth station that had been suggested by the IDC could not accept Intelsat signals. It was a medium-sized earth station that was expected to be erected in remote stations and settlements, and was known as HACBSS (Homestead and Community Broadcast Satellite Service). The price of the ABC's participation in Aussat was $38 million over the seven-year anticipated operating lifetime of the first generation satellites. This was money that had seldom been harder for the organisation to find, as it had suffered cuts to its budget under the Fraser government, a process that continued after the 1983 Hawke election.

On its election in March of that year the Australian Labour Party (ALP) was confronted with an immediate problem about Aussat. Having opposed the project at various stages of its development, it had a limited range of tough options to exercise, and one of them was easily the most politically expedient. The toughest option, and the one most consistent with its policy when in Opposition, was to abandon the satellite project entirely. That was considered too politically difficult by the new Government, especially as such a move could be represented as a blow aimed against rural voters, who traditionally supported the conservative coalition Labour had tipped from office.[7]

201

The sentiment that political prudence demanded the satellite project should proceed was reinforced by the putative cost of its cancellation. The federal Department of Communications, which had been able to extend its influence and expand its size on the strength of its co-ordination of the satellite project under Ministers of the previous government, strongly advised against cancellation. It was responsible for putting into circulation a figure of $150 million as the potential loss in the event of cancellation. That figure was accepted within Government circles, despite the questionable premises upon which it was based. Instead of simply withdrawing from its contracts and paying penalty clauses, which may have incurred the Government substantial losses, the ALP had other options.

One which was barely explored was the possibility of reconfiguring the Aussat satellites under construction by Hughes in Southern California to render them attractive commercial propositions. The Hughes HS 342 satellite series was the most commercially successful so far devised. With relatively minor modifications, Aussat's versions of this series could have been offered for sale to telecommunications carriers on the threshold of rapid expansion following the announcement of the divestiture of A T & T in January 1982. It could also have been an attractive purchase for companies who wanted it for television distribution, especially for direct broadcast. In addition, there was no shortage of corporations planning telecommunications networks to by-pass the Long Lines Division of a newly deregulated and restructured Bell System. But the prospect of undertaking the commercial task of testing the market for potential purchasers proved too daunting for a new Government under pressure from media companies and its own federal bureaucracy to proceed with the satellite system without delay.

Short of abandoning the project entirely, the new government could have ensured that the satellite system proceed in a form that met objectives that differed from those of its predecessors. During the process of its planning over the previous six years, two major issues about Aussat had attracted controversy. The first was the possibility that Aussat's high-powered transponders would be purchased by the three largest metropolitan television networks. This would enable the big media owners to attack the markets of the monopoly regional commercial broadcasters, bringing additional revenue when the growth of metropolitan advertising rates had slowed. Impetus for this use of the satellite increased from 1983 when it was clear that metropolitan television advertising rates had peaked. Their networked signals beamed across the continent would ensure that remote rural dwellers, equipped with their own satellite dishes, could receive programmes. But so could people who lived just out of reach of the signal range from the major metropolitan stations. The more densely populated rural areas of the two largest states, New South Wales and Victoria, represented potentially lucrative new territory for the networkers. This regional market consisted of 1.75 million viewers, an audience which if it could be reached by satellite represented a considerable revenue boost even to the big city networks.

The second major objective of the satellite system's promoters was the deregulation of telecommunications by a strategy in which Aussat played a central role. Under the Fraser government it had been proposed that Aussat, with a minority government interest, should be permitted to compete with the national monopoly telecommunications carrier, Telecom Australia. By competing for the carriage of corporate data traffic on the eastern seaboard, Aussat was expected not only to push Telecom tariffs down for business services, but to siphon off revenue from the national carrier by offering cheaper rates. The effect would be a levering open of Telecom's monopoly and the entry of private sector carrier competition. The models for such a strategy were the closely observed activities in North America of IBM's Satellite Business Systems, and MCI Communications, and in Britain the Mercury consortium. As A T & T was obliged to begin restructuring and British Telecom was privatised and opened to competition, proposals were made by large telecommunications users to fragment Telecom Australia.[8]

Both the television networking and telecommunications deregulatory objectives of the Fraser government were likely to be changed by the incoming Labour government. It was a matter of deciding how great were to be the amendments. In television broadcasting, the government began a series of policy-making reports designed to maximise the political advantages of the inherited satellite system. Instead of national networking remaining at the top of the agenda, the possibility of country television viewers having a selection of channels became the central theme of the government's deliberations. In telecommunications, the new Minister of Communications, Melbourne lawyer Michael Duffy, personally favoured making Telecom the satellite system's owners. This would have ensured its technical integration with the national terrestrial network and have thwarted Aussat's competitive ambitions. Duffy lost the decision in Cabinet to more senior Ministers with less desire to protect Telecom from competition. And a vote by the full parliamentary Labour party that would have forced Cabinet to transfer ownership to Telecom, was lost by a single vote. The government took six months to determine Aussat's new direction. In November 1983 it announced that its predecessor's ownership plans had been changed. No longer was there a 49 percent/51 percent government/private sector investment mix. Seventy-five percent of Aussat's equity was to be owned by the government, with the remaining 25 percent allocated to Telecom, which had the right to nominate two members to the board of directors. The representatives of large companies and the main media proprietors who had collectively dominated the board under the Fraser government, were mostly replaced.

Meanwhile the government continued its attempts to devise the most politically acceptable arrangement to increase the number of television channels rural dwellers could receive. They culminated in the last-minute passage of legislation through the federal upper house, the Senate, on the eve of the 1987 election. With the return of a Labour government at the July 11 election, that legislation has become the blueprint for the pattern of television broadcasting by satellite until at least the mid-1990s.

The federal Department of Communications had identified two major options facing the government if it was to achieve its aim of providing more television broadcasting in rural Australia. The most obvious was to open all the existing regional markets to the degree of competition that existed in the major cities. In addition to the ABC and the Special Broadcasting Service (SBS), the national ethnic television broadcaster, there would be three commercial licences offered. While this would have initially satisfied the demand for additional services, there were serious impediments to the adoption of this straightforward proposal.

Although regional commercial broadcasters enjoyed monopoly rights in their own areas, their markets were very small, in some cases consisting of a few hundred thousand people. The prospect of dividing relatively meagre revenues three times would have daunted potential commercial broadcasters from entering the market. And where it did not, it was difficult to see how both the new entrants and the one existing broadcaster could make sufficient returns on their investments to sustain their operations. The survivors of such intense competition in these small markets were likely to be those operators that could cut their operating costs below that of their competitors. At best, the most costly components of regional broadcasting, the making of local news and current affairs, would be threatened by cut-backs or elimination. At worst, network affiliates receiving programme feeds from Sydney or Melbourne and re-transmitting them regionally, would push the existing broadcasters, saddled with much higher operating costs, into liquidation.

This was an unacceptable political consequence for the government because it would allow its policies to be represented in rural electorates as inimical to the interests of country dwellers. The political impact of the closure of regional media companies would have been especially damaging in a period when country areas were just recovering from severe drought in the early 1980s. Depressed world commodity prices from 1984 onwards had created a sense of political crisis in rural Australia, leading to unprecedented militancy among farmers. The share holdings in many of the local media companies were widely spread among the small business operators, lawyers, doctors and the wealthier citizens of country towns. The prospect of the threatened closure of 'local' television channels and their replacement with networks broadcasting from the large cities would have been a tempting rallying point for a formidable coalition of rural interest groups. The government's desire to protect the parochial flavour of regional broadcasting became enshrined as doctrine in the policy-making process, where it was described as 'localism'.

The metropolitan-based networks had, as early as 1977 when Kerry Packer commissioned Donald Bond's consultancy, targetted regional areas for expansion. The availability of a domestic communications satellite was central to those ambitions. Donald Bond wrote his report just as discussion about direct broadcast satellites began to interest television companies and media entrepreneurs in North America. But his proposal for an Australian communications satellite was much less 'state of the art'.

He had envisaged the city-based networks extending their broadcasting range into the continent's hinterland, where signals in C band would be picked up by relatively large earth stations and conventional terrestrial relays used to reach viewers. The cost of the technological infrastructure necessary for the networks' expansion would mostly be borne by the operator of the satellite system. The networks' costs would be quickly recouped if their strategy of crushing competition from existing local broadcasters succeeded. It was never Packer's original intention to own and operate the satellite system himself, although he was later to offer to do so as a gesture of frustration with the length of time it took the government to act on his proposal. The relatively low entry cost into regional television markets by the metropolitan networks, their access to more recent and higher rating programmes, and their general marketing superiority, understandably intimidated rural broadcasters. The programme costs incurred by the networks would already have been more than met by revenues from their metropolitan advertisers and the new regional markets provided the prospect of almost total profit.

From the time it became clear that the Fraser government was taking the Packer satellite proposal seriously, regional broadcasters began to point out the vulnerability of their advertising base to network competition. The lobbying power of rural television operators was used to alter the shape of plans by Fraser's Minister, Tony Staley, to issue supplementary television broadcasting licences in country areas. The lobbyists based arguments about their vulnerability on the fact that about 70 percent of their advertising revenue was derived from clients who advertised on a national basis. That revenue could be heavily eroded by advertising packages put together by network television broadcasters for national clients who had hitherto used regional broadcasters to extend their coverage beyond the cities. The residual local advertising market was insufficient to keep them in business, argued the rural television broadcasters.

The arguments of the rural television proprietors that their existence be guaranteed under a competitive regime was quickly accepted by both sides of Australian politics. When the Hawke government took office in March 1983 it proposed different arrangements for competition among commercial broadcasters in rural areas, but continued to recognise that the viability of the incumbents should not be threatened. But that did not mean those operators were to be left free to continue making the large profits that had been a feature of their balance sheets since the late 1970s. In the ten years to 1984 regional broadcasters had experienced an average revenue growth in real terms of 12 percent per year.[9] The new government made clear that when competition was introduced it had to be managed to ensure the survival of all the commercial broadcasters. It took more than three years to find a means of maintaining the delicate balance between giving the networks access to the country, and protecting the existence of the rural broadcasters.

The term that the Department of Communications coined for the regulatory device intended to achieve this complex aim was 'aggregation'. Its approach involved looking at regional television as a series of markets each of which was by itself too small to sustain three commercial competitors. It concluded that any solution could only be based on expanding the size of those markets. The Department proposed that neighbouring regional television broadcasting areas should be aggregated into markets of about a million viewers each. That was the minimum size considered large enough to sustain the existing broadcasters and two new competitors. A timetable was set by which this policy of aggregation would apply to the southern area of New South Wales, to neighbouring Victoria by 1990 and progressively to the other states and the Northern Territory by 1996. In all, six new aggregated regional markets were to be created. Areas that did not geographically lend themselves to this process, or which were too small, were excluded.

While a formula for balancing the conflicting interests of regional and metropolitan television broadcasters was being prepared, the purchase by the major networks of transponders on Aussat began to force to the surface a potentially even more contentious issue. This was the predictable desire of the networks to rid themselves of government regulation that restricted their ownership to no more than two major city television stations. The availability of transponders on the domestic satellite, and the depressed state of the television advertising market, created pressures that were to result in the rewriting of government policy. The emergence of three large national commercial networks using satellite feeds to deliver programmes to affiliates began to be treated as a serious possibility.

By the election day of July 11, 1987 the entire ownership pattern of the Australian media had been transformed. With hindsight, it is possible to nominate when the starting gun was fired to begin a scramble by some of Australia's largest corporations to divest and acquire media interests. Communications minister Michael Duffy had pressed the trigger on November 27, 1986 when he announced that existing prohibitions on television broadcasting ownership were to be abandoned. Until then, no single proprietor could concurrently own more than two television stations in major cities. Broadcasters were not prohibited from owning other media

interests and the three largest groups were both newspaper
proprietors and owners of metropolitan-based television channels.
From the date of his announcement the two station rule was to be
replaced by a proposed limitation on any single proprietor having
access to more than 75 percent of the total viewers in Australia.

In return for creating the conditions by which the major
metropolitan television station owners might fulfil their networking
ambitions, the Hawke government introduced the first
cross-ownership restrictions ever applied to the Australian media.
Under the terms of the government's new policy, no single owner
could both operate a television channel and publish a major daily
newspaper in the same city. The policy did not disturb existing
cross-ownerships of that sort, exercised in Sydney by the Fairfax
group and in Melbourne by the Herald and Weekly Times. But any
changes of ownership after November 27, 1986 would be caught by
the new provisions.

The 75 percent cut-off was regarded as far too high by the
regional television broadcasters; their lobby group had proposed a
43 percent maximum, equivalent to a channel in each of the two
largest cities. Public interest groups, who made presentations to the
inquiries that preceded the government's decision, suggested
figures of less than half that. The percentage eventually adopted in
June 1987 was 60 percent, a percentage struck after hurried
pre-election negotiations with Opposition groups in the Senate. The
balance of power in the Senate was held by the Australian
Democrats, a small party occupying the increasingly difficult to
identify middle ground between Labour and the Liberal/National
party coalition. Together with an even smaller number of
independent Senators, the Democrats had been able to exercise a
veto over government proposals to trade off the two-station rule for
cross-ownership prohibitions. The Democrats' resistance was
based on the argument that the new policy would encourage
concentration of ownership, albeit in different media. For while the
government's proposal permitted the creation of more extensive
television networks it would also encourage concentration of
ownership in the print media by companies that were not television
broadcasters.

The last-minute agreement with National Party Senators resulted
in the government reducing the marketing ceiling for television
networks by 15 percent, a cut that would not seriously inhibit the
creation of very much bigger networks than had previously been
permitted. The conservative Nationals, with an electoral base in

rural Australia, were nervous of going into an election in which they could be represented by the government as having denied additional television services to country viewers. The party had, under the stress of its own faction fights in the preceding months, declared an end to the formal coalition arrangement it had with its conservative allies, the Liberals. This allowed National Senators to vote with the government to pass legislation confining ownership of a single television proprietor to 60 percent of the total market. By negotiating the government down from 75 to 60 percent the Nationals placated the rural television proprietors at the threshold of an election campaign in which political coverage on television was increasingly important.

Meanwhile, the changes that Duffy announced to the ownership provisions of the Broadcasting Act had been the beginning of the most extraordinary three months in the sometimes turbulent history of the Australian media industry. Between November 1986 and February 1987, of the six commercial television stations in Sydney and Melbourne, five changed hands. Changes of similar magnitude occurred in other cities and towns around Australia as a bewildering series of transactions unfolded.[10]

The most straightforward of the ownership changes occurred in Sydney and Melbourne, the two largest markets, and involved the transfer of the two most profitable commercial channels in those cities. The seller was Kerry Packer, whose family company controlled the Nine network stations in Sydney and Melbourne, a string of metropolitan and country radio stations, a stable of large circulation magazines, and diversified resources and property holdings. The buyer was the takeover specialist Alan Bond, who had built an empire based on resources and property and added major breweries and new ventures such as the development and manufacture of airships for military use. The deal was struck at a little under a billion dollars and gave Alan Bond a network of four metropolitan television stations, embracing Sydney, Melbourne, Brisbane and Perth. Like Packer, he had no metropolitan newspaper operations and was exempt from any possibility of divestiture because the market he reached did not exceed the government's proposed 75 percent of the total.

Bond's new network of radio stations included one in Sydney, later to be sold off, others in Melbourne, Perth and Darwin, and five regional radio stations in rural Western Australia. The technology to knit these acquisitions together was also purchased from Packer, a 30 watt transponder on Aussat. The deal included Packer's lease of Intelsat satellite circuits that permitted the Nine network to directly import programmes from the major networks in the USA and to broadcast them live. Packer had used this link since 1982 when he constructed an earth station at his Sydney studios, by-passing the facilities of the Australian international traffic carrier (and Intelsat member), OTC. It enabled the Nine network, later emulated by the Seven network, to broadcast American news and current affairs programmes live between midnight to dawn in Australia.

Bond's Sky Channel service, which broadcast sporting programmes direct to clubs and hotels using Aussat, fitted neatly into the mosaic of his purchases from Packer. Faced with dwindling patronage because of changed drinking habits, tougher drink-driving laws and competition from more sophisticated entertainment venues, many clubs were winning customers for Sky Channel programmes. For the price of erecting a satellite dish on their roofs, or in their parking lots, they were able to advertise themselves as having joined the satellite age. Live broadcasts of events such as world title boxing matches, international football games and other sporting events, were presented as an antidote to falling club revenues. In most cases, the broadcasts were exclusive to Sky Channel, obliging patrons who were keen followers of particular sports to watch major events in the clubs. Bond marketed Sky Channel services by packaging them with sales of his brewing products. Clubs which undertook to purchase a certain volume of liquor could acquire the Sky Channel service at discounted rates. This practice was later abandoned by Bond in the face of questions about its status under trade practices legislation.

While Packer's divestiture of the Nine network occurred swiftly and privately as a result of personal negotiations with Bond, the redistribution of the broadcasting assets of Australia's largest media group, the Herald and Weekly Times was complex and public. The Melbourne-based company had major newspapers in every state, and others were published in Papua New Guinea and Fiji. It also published many smaller local newspapers and a collection of rural newspapers and specialist magazines. In television, it formed the Melbourne axis of the Seven network in which its partner was the Sydney-based Fairfax media group. In radio, it owned metropolitan and rural radio stations and it had set up companies in developing areas such as electronic information services. A long-established company, as conservative in business practice as it was in editorial outlook, the Herald empire had thwarted its most ardent suitor in 1979 when it defeated, with assistance from Fairfax, a takeover bid from Rupert Murdoch's News International.

In the intervening years it was periodically subject to speculation that another takeover attempt was about to be launched. The Perth-based Robert Holmes a'Court had used share raids to assemble an embryonic media empire in Australia, and another in Britain. His steady acquisition of shares in the publicly-listed Herald group was interpreted by the target company's board of directors as the construction of a launching pad for a takeover bid. When the battle for the Herald empire blazed into the open with a formal offer for its shares by Murdoch, the response from Holmes a'Court was swift. His counter-offer made clear that whoever proved the winner, the contest was to be fought not only on the basis of price per share but would also depend on which suitor won a recommendation of acceptance from the Herald's board. The frenzied auction brought the Fairfax group into the fray, with its main concern the probability that the Seven network's Melbourne television channel might fall into the hands of a rival.

The recommendation by the Herald's board that the Murdoch offer be accepted meant that control passed to him, but it also posed the question of how the resulting complex sell-off of assets would be conducted in order to meet the restrictions on cross-media ownership imposed by the government. Under the government's new guidelines, Murdoch would have to divest himself either of major metropolitan newspapers or the television stations he already owned plus those he had acquired with the purchase of the Herald and Weekly Times. Although both Holmes a'Court and the Fairfax group had failed in their bids to gain control of the Herald group, they proved willing purchasers in the forced sale of assets that followed Murdoch's victory. Even before negotiations had been concluded with the Herald group over what constituted a price its board could accept, discussions with potential customers for fragments of the media empire had begun. So skilfully had the takeover been pitched that by the time the Herald's acquisition was complete almost all the $3.4 billion that the Murdoch family company, Cruden Investments, had committed to the takeover had been recouped from assets sales. Murdoch was to effectively pick the eyes out of Australia's largest media empire, retaining what he wanted and discarding what he didn't, for a net outlay of zero.

Murdoch's first set of negotiations were designed to find a purchaser for his Channel Ten network stations in Sydney and

Melbourne. They were the third-ranked of the three capital city networks in revenue terms but in Murdoch's hands they proved valuable assets. They were purchased by a consortium of an existing broadcaster and a property developer for a price of $842 million. The buyers were the North Star media group, whose base of operation was in newspapers, television and radio stations in Northern New South Wales and Southern Queensland. Its partner was Westfield Capital Corporation, a new entrant into the ranks of media owners, and a spin off from a company that had established its fortune by constructing giant shopping centres. The North Star/Westfield network created as a result consisted of radio and television outlets in Sydney and Melbourne, two more television and two radio stations in northern NSW, and three Queensland radio stations, including one in Brisbane. With the addition of the two former Murdoch television stations North Star/Westfield owned a network that had access to 51 percent of the total Australian market. Although that figure exceeded the 43 percent limit urged on the government by regional broadcasters, it fell well short of the government's proposed ceiling of 75 percent.

Holmes a'Court, defeated by Murdoch in the fight for the Herald group, was consoled by the creation of a television network that linked a station in Perth to another in Adelaide. He added to that four radio stations, one of them in Perth, plus the largest circulation daily newspaper in Western Australia, another acquisition from Murdoch's asset sale of the Herald group. By owning a major metropolitan newspaper and a television station in Perth, Holmes a'Court put himself in breach of the government's cross-ownership prohibition. When legislation passed in the dying days of the second Hawke government is promulgated, the enforcement role will rest with the regulatory watchdog, the Australian Broadcasting Tribunal.

The Tribunal's influence in policy-making had proved minimal, with the clearest example of the government's attitude to its advice coming in 1984 when it prepared a report on new ownership limits in television broadcasting. Its proposed ceiling of one television licence per major city amounted to a maximum of about 25 percent and would have prevented the creation of further networks to rival the audience size enjoyed by the three existing networks with channels in Sydney and Melbourne. The Tribunal's role was confined to issuing supplementary licences in rural areas, where the successful applicants ranged from an Aboriginal Broadcasting Group, to a consortium of all the major regional broadcasters.

Another Perth-based entrepreneur, Kerry Stokes, already an owner of television and radio stations, proved a willing buyer of other Herald assets disposed of by Murdoch. By purchasing the Herald group's Adelaide television station and its Melbourne radio station Stokes extended his existing network to embrace three major cities and some large rural ones. As a result of his Murdoch purchases he now owned television channels in Canberra, Perth and Adelaide and radio stations in Perth, Adelaide and two large provincial centres, one in Victoria and another in South Australia.

The Fairfax group, with most to lose from Murdoch's Herald takeover, was obliged to pay expensively to retain its Seven network. It also confronted the divestiture demand that confronted Robert Holmes a'Court because the purchase of the Herald's HSV 7 television station in Melbourne conflicted with its ownership of *The Age*, the largest circulation quality newspaper published in that city. The Fairfax group had only two years previously completed the acquisition of all the shares in the publisher of *The Age*, David Syme and Co. In Sydney, the largest single television market, Fairfax had for many years owned both *The Sydney Morning Herald* and ATN 7 television channel, and the group was unaffected by the government's new legislation. Fairfax had paid Murdoch $320 million for the Herald's loss-making television station in Melbourne and while it signalled an intention to stave off divestiture either of its new asset or one of its most profitable newspapers, it sought to reduce operating costs. A series of industrial disputes involving production staff employed by HSV 7 followed Fairfax's takeover as the new owners attempted to cut local programme-making and met resistance. Ratings for the station's evening news broadcast, the key to attracting viewers whose inertia would keep them tuned to a single channel for the evening, plummetted. In public discussion in the letter columns of newspapers and the talk-back programmes of radio stations, the Fairfax group was represented as a Sydney-based interloper, determined to rid its television outlet of any involvement from Melbourne. Sensitivity to this possibility had been heightened by the progressive down-grading of Melbourne's TCN 9 television station under the ownership of Kerry Packer. In the year preceding his sale of the station to Alan Bond, the total number of jobs lost had exceeded 200; the cut-backs were blamed on advertisers unwillingness to pay increased rates.

The dramatic nature of the wholesale shifts in ownership of television, radio stations, newspapers and magazines, which presented a rare spectacle in Australia of the media covering itself, obscured the acceptance of 'equalisation' in rural television. When the government had put its amending legislation to 'equalise' television in the country by aggregating regional markets before the Senate in December 1986 it was stalled by the establishment of a Senate Select Committee. The balance of power in the upper house was exercised by the Democrats and Independents to call a summer moratorium on the government's proposals. The Select Committee became a forum in which arguments supporting and opposing the government's plans were canvassed.

By the time the Select Committee had finished its deliberations early in 1987, the government had made its intentions clear. But the Committee's hearings had proven one of the few public forums available for discussion about the implications of the turmoil of the preceding few years. A number of public interest groups made submissions to the committee and it was in their contributions — reflected in dissenting reports in its final report — that the broader issues were canvassed. Neither the prolonged policy-making process nor the spectacular examples of the free play of market forces, had addressed their concerns. Why was there an assumption, they asked, that increased commercial television services represented 'an increase in choice and diversity in the broadcasting system'.[11]

The public interest groups' response to the wholesale re-shaping of television ownership sparked by the availability of satellite transponders was to question how the process had done anything but entrench the power of the metropolitan networks. They pointed out that it was the small size of the total Australian television market that restricted quality programme-making, and that was unlikely to be reversed by the emergence of new broadcasters. The consequence of all the developments since 1977 was that the creation of three major US-style commercial television networks seemed inevitable. And that of course was very much the model for Kerry Packer's original intervention that became the first step in a process that led to Aussat.

In the decade of Australian satellite broadcasting policy development and implementation there were many situations that in the midst of uncertainty and political horse-trading retained their irony. None was greater than the decision by Kerry Packer to sell the two most valuable television channels in the land, just as his original network idea began to be realised. What is more, he had accepted an offer of a billion dollars to sit on the sidelines and leave to others the task of judging whether it was a good idea or a bad idea. In the early stages of the July 1987 federal election, the billionaire former television proprietor expressed a political preference that was reported in newspaper headlines. He had, he declared, decided to cast his ballot for Prime Minister Bob Hawke's Labour Party.

REFERENCES

1. The development of Canadian satellites is put in an international perspective by William H. Melody, 'Direct Broadcast Satellites; the Canadian Experience', *Symposium on Satellite Communication; National Media Systems and International Communication Policy*, Hans Bredow Institute for Radio and Television, University of Hamburg, December 11, 1982.

2. The definitive account of the history of Australian telecommunications is Ann Moyal's, *Clear Across Australia*, Nelson, Melbourne, 1983.

3. For an account of the political process leading to the decision to launch Aussat, see Ian Reinecke and Julianne Schultz, *The Phone Book*, Penguin, Ringwood, 1983, especially chapter six.

4. This possibility is further explored in Ian Reinecke, 'Life Without...We Could Have Said No', *Media Information Australia*, no. 38, November 1985.

5. See Trevor Barr, *Electronic Estate*, Penguin, Ringwood, 1985, provides a blow by blow account of the decision-making process in chapters seven and eight.

6. Ian Reinecke, *Connecting You*, McPhee Gribble/Penguin, Ringwood, 1985.

7. For a discussion of alternative uses of Aussat, see Ian Reinecke's *Connecting You*, 1985.

8. An account of the first generation uses of Aussat are collected in *Aussat '86: New Horizons*, Conference Papers, Aussat Pty Ltd, Sydney, November 1986.

9. Trevor Barr, *Electronic Estate*, Penguin, Ringwood, 1985.

10. The account of detailed changes of ownership is drawn mainly from *The Sydney Morning Herald*. Like the other major metropolitan newspapers, it carried almost daily accounts of developments in the media over this period. For a useful summary, see 'The Great Media Carve-Up', *The Sydney Morning Herald*, February 10, 1987.

11. *Submission to the Senate Select Committee on Television Equalisation*, Academics for Responsibility in the Media, Canberra, January 1987.

8

Satellite Broadcasting in the United States

Heather E. Hudson

EARLY HISTORY

Arthur Clarke's article in *Wireless World* in 1945 proposing a 'space station' marked the beginning to the satellite era. However, until the shock of Sputnik in 1957, U.S. space research including work on telecommunications systems was carried out primarily for the military and was virtually invisible to the public. However, after Sputnik, the U.S. made an accelerated commitment to its civilian space programme, highlighted by the establishment of the National Aeronautics and Space Administration (NASA) in 1958. In December, 1958, the Department of Defense launched its first communications satellite, Signal Communications Orbiting Relay Experiment (SCORE).

In 1961, President Kennedy set the goal of sending a man to the moon by the end of the decade, and promised a major commitment to space research and applications, including the implementation of a communications satellite system. Government and the private sector co-operated to implement this goal. In 1962, NASA launched Telstar, a low altitude satellite designed by Bell Labs and built by A T & T. RELAY, another low altitude satellite built by RCA, was launched in December 1982. The first geosynchronous satellite, Telstar, built by Hughes Aircraft under contract with NASA, was launched in 1963.

The International Communications Satellite Organization (INTELSAT) was established in 1964, largely at the impetus of the United States (see below). Its first satellite, INTELSAT I, known as EARLYBIRD, (also built by Hughes) was launched by NASA in 1965. Thus commercial satellite service became a reality less than five years after President Kennedy's commitment to space and only two years after the first experimental geosynchronous satellite had been launched.

THE EXPERIMENTAL ERA

In 1966, NASA launched the first of its Applied Technology Satellite series, ATS-1. These satellites were built for technical experimentation to test and evaluate new technologies. ATS-1 was designed with a two year life; by 1970, it was still functioning. The state of Alaska then requested to conduct experiments with ATS-1 to determine whether satellite communication would be a means of improving telecommunication services to the bush. Like northern Canada, Alaska had very poor communications, primarily by high frequency radio which was subject to outages depending on the weather, the season, and the sunspot cycle. In addition, most of the radios in villages were owned by a variety of government agencies and private organisations such as church missions and merchants, so that there was no publicly accessible telecommunications system. Also, many of the sets were not well maintained.

The National Library of Medicine supported a project on ATS-1 to determine whether satellite communications could improve the quality of health care of Alaska natives. Satellite earth stations were installed in regional hospitals and the Alaskan Native Medical Center in Anchorage, and in villages in the Tanana region of central Alaska. A daily 'doctor call' by satellite linked the doctors at the regional hospital to village health aides who received advice on the diagnosis and treatment of their patients. The single channel conference circuit turned out to be a valuable instructional tool: by listening in to 'doctor call' the health aides learned new ways to treat their patients. The satellite was also used in emergencies to arrange for evacuations by plane. The project was highly successful: the satellite technology was much more reliable than high frequency radio, and in the first year the number of patients in the satellite villages treated with a doctor's advice more than tripled.[1]

The State also supported an education project to link village schools and to produce radio programmes about village life. Microphones and speakers in the classroom allowed students to talk to their counterparts in other villages. Children and adults came to the schools in the evenings to hear programmes about native topics including whaling in Barrow, a stick dancing festival, Arctic winter games, keeping healthy dogs, and native land claims.

The ATS-1 satellite was situated far enough west to cover all of Alaska as well as the continental U.S. As a result of its position and its global beam, its signal also covered much of the Pacific. In 1972, the University of Hawaii started a project known as PEACESAT that established a network of home made ATS-1 terminals throughout the South Pacific and as far as Australia and New Zealand. The University of the South Pacific, based in Fiji, also set up USPNET to link its extension centres in island nations across the South Pacific. Many of its centres also participated in PEACESAT. Over the years until the demise of ATS-1 in 1985 (nearly 20 years after its launch) there were several exchanges between the South Pacific and Alaska, and occasionally physicians in the South Pacific gave advice to Alaskan health aides!

ATS-1 was followed by several other experimental satellites. The ATS-3 satellite was also used for educational projects in the continental U.S., Alaska, and the Caribbean. After the moonwalk in 1969, public pressure increased to apply the advances of space technology to problems on earth. One of the results was that future NASA satellites were planned to include social experiments. The ATS-6 satellite was used for several medical and educational experiments. In Alaska, the Public Health Service experimented with 'telemedicine' including video links between two village clinics, a regional hospital and the Alaska Native Medical Center. Patients could be examined remotely from the village, and training materials could be transmitted over the network. The evaluators found that although useful in most cases, the video was not cost effective. Given the limited facilities in the villages, there was little that could be done for patients without evacuation, and reliable audio communication seemed adequate to assist the health aides in diagnosis and treatment. However, consultations from the regional hospital to the medical centre seemed to benefit more from the video link.[2] Video clinics were also used for consultations with physicians in Seattle and for links between the medical school at the University of Washington and pre-medical students not only in Alaska but also in Montana and Idaho.

A second cluster of projects involved enrichment of education for programmes to rural schools in the Rocky Mountain region. Instructional materials produced in Denver were delivered to selected schools by satellite. A package of existing educational materials appropriate for courses identified as priorities by teachers were also delivered by satellite. There were also teleconferences between students, as well as for teachers and parents using interactive audio.

A third experimental region was Appalachia, where both medical and educational experiments attempted to overcome the isolation of small scattered communities. One project delivered continuing education programmes to teachers, nurses and others who did not have access to specialised career training. Programmes were delivered to community colleges and fed into local cable systems. The project was continued on the CTS satellite (see below) and then ACSN was launched as a service to cable networks with a transponder on the commercial RCA SATCOM satellite. ACSN's directors planned to make it easy for cable operators to pick up their programmes by using the same satellite that delivered Home Box Office. Their constituency expanded beyond Appalachia to include the whole nation. Eventually, they changed the name of the network to the Learning Channel. ACSN/The Learning Channel is clearly one of the success stories of the experimental era.

The last of the experimental satellites was the Communications Technology Satellite or CTS, known in Canada as HERMES. This satellite was jointly sponsored by NASA and the Canadian Department of Communications. Again, it included telemedicine and tele-education experiments, some of them continuations or refinements of experiments begun on CTS. There were also a few experiments linking organisations in both countries such as an exchange of engineering courses between Stanford University in California and Carleton University in Ottawa.

By the mid 1970s it was decided that NASA had fulfilled its experimental mission in satellite communications, and that the technology had now been successfully transferred to the private sector. A tiny budget was reserved to keep the satellites operating. ATS-1 and ATS-3 in particular refused to quit, and a lobby from the Pacific reversed NASA's plans to turn them off and remove them from orbit. It is interesting to note that at present NASA is back in the experimental satellite field, with plans for a high powered Ka-band satellite (Advanced Communications Technology Satellite or ACTS) to be launched in 1990. Industry led the lobby for renewed NASA involvement, stating that the private sector would not be able to invest in the necessary research and development to field test technology using frequencies above 20 GHz. (Meanwhile, Japan is building a Ka-band satellite, and

Hughes Aircraft has announced that it will also offer commercial satellites using these frequencies.)

OPEN SKIES

The advent of satellites posed major policy dilemmas for the U.S. In the early 1960s, policy makers were divided on whether satellites represented a revolutionary new technology that would change the face of telecommunications or a supplementary technology to existing terrestrial facilities. There was also disagreement on how the technology should be owned, operated and applied. Should it be operated by the common carriers as advocated by the Federal Communications Commission (FCC) and the carriers? Should it be publicly funded but privately owned and operated for the benefit of all mankind as advocated by NASA spokesmen? Or should satellites be operated as a government-owned enterprise as advocated by some liberal members of Congress?[3]

International decisions came first. In 1962 Congress enacted the Communication Satellite Act which established COMSAT as the 'chosen instrument' to 'establish as expeditiously as practicable a commercial communications satellite system as part of an improved global communications network' and to 'direct care and attention toward providing ... services to economically less developed countries and areas as well as those more highly developed.'[4] COMSAT was to be half owned by the U.S. international carriers and half owned by the general public. In 1964, INTELSAT (the International Telecommunications Satellite Organization) was established, with COMSAT acting as the U.S. representative. Before long international communication via satellite became a reality.

In April 1965, NASA launched the Early Bird satellite (also known as INTELSAT I) which inaugurated commercial international satellite services. Five months later, the American Broadcasting Company applied to the FCC to establish a domestic broadcast distribution service via satellite. However, opinions were much more divided concerning domestic services. After six months of deliberation, the FCC returned the application without prejudice because of the novel legal and political questions it raised, and initiated instead formal hearings to determine whether the FCC could inaugurate such systems, and if so under what terms and conditions. Four entities then proposed domestic systems:

ABC, A T & T, COMSAT, and the Ford Foundation. In a related development, Congress established the Carnegie Commission on Educational Broadcasting which proposed in 1967 the establishment of a Corporation for Public Broadcasting and recommended preferential treatment for public broadcasting when satellite distribution services became available.

The complexity of these issues and the conflicting views being advocated on both domestic and international policy issues led President Johnson to establish a task force to recommend a national telecommunications policy to be chaired by Undersecretary of State for Political Affairs Eugene Rostow. In 1968, before the Rostow Task Force issued its final report, President Johnson announced his intention not to seek re-election. When the final report was made public by the Nixon administration, it carried no formal endorsement by either the Nixon or Johnson administrations.

However, domestic satellite policy developed not from the Rostow Task Force recommendations (which favoured a government operated monopoly) but from the White House itself. During the Nixon administration a powerful Office of Telecommunications Policy was established in the White House under director Clay Whitehead. One month before OTP was established, a letter from the White House to the FCC stated that the government should encourage and facilitate the development of commercial satellite communications to the extent that industry found them economical and operationally feasible. Subject to technical requirements to prevent harmful interference and anticompetitive practices, any financially qualified entity, private or public, should be permitted to operate domestic satellite facilities. In March 1970, the FCC issued a report and order outlining a policy of open entry which in a 1972 Memorandum Opinion and Order delineated what became known as the 'Open Skies' policy.

EARLY SERVICES

The first domestic satellites to be launched were SATCOM I owned by RCA and WESTAR I owned by Western Union. ATT was barred from operating its own satellite systems for seven years lest it dominate the entire industry; however, it participated in the COMSTAR system operated by COMSAT.

The satellite operators expected that the television networks would be their first major customers. However, the networks already had long term contracts with A T & T for microwave links to their affiliates and were reluctant to abandon the proven system for the new technology. Also, it was not clear who would pay for the earth stations if satellites were used; the network did not want to instal earth stations at each affiliate, and the affiliates were unwilling to bear the cost. (This issue remained unresolved until the 1980s, by which time the cost of earth station technology decreased and the networks were willing to enter into agreements with the affiliates for co-operative financing of the facilities.) Instead, a pay television service called Home Box Office (HBO) became the most successful customer. HBO had begun to offer movies to cable systems locally and through limited microwave distribution. HBO officials saw satellites as a means of reaching every cable system in the country. And pay television was the attraction that cable needed to attract urban customers who already had several over the air programme choices.

Satellite-delivered pay TV and cable turned out to be a synergistic combination that fuelled the growth of the cable industry and its penetration of the major urban markets that had recently been opened by the FCC for cable franchising. Thus, for the next decade, the major customer of domestic satellites was cable television networks. Voice, data, interactive video teleconferencing and specialised video services grew slowly and have more recently become established.

CURRENT STATUS OF THE SATELLITE INDUSTRY

As of March 1986, eight companies owned and/or operated 17 C-band (6/4GHz) (3 co-located), 7 Ku-band (14/11GHz), and 3 hybrid commercial U.S. domestic satellites. This list does not include companies that lease transponder capacity in bulk or that own one or more transponders. Approximately 40% of the transponders in use, both C-band and Ku-band, are devoted to satellite television service. The majority of these transponders are for distribution of channels to cable systems.

Satellites have transformed the cable industry in the U.S. from 'mom and pop' operations bringing in distant over the air signals to a major business with annual revenues of $10 billion. There are now over 37 million cable subscribers in the U.S. out of a total of 87 million television households. The availability of non-broadcast signals has transformed cable to a primarily urban enterprise.

The networks have also shifted their delivery of programming to affiliates from microwave to satellite, and they also use satellites for programme feeds, remote coverage, and special coverage. Programme syndicators and specialised and regional networks also use satellite transmission. Private television networks for in-house teleconferencing and occasional teleconferences also contribute to video capacity. In many cases, these latter services use only limited time periods on a transponder and may be shared with other users, such as cable channels that broadcast for a few hours per day. The satellite brokers find time for these occasional users.

The following sections describe the various services provided by U.S. commercial satellites.

Cable programming via satellite

The major types of programming delivered via satellite to cable subscribers are:

Pay Television: movies are the major pay service for which viewers pay a surcharge each month in addition to monthly cable charges. The major channels are Home Box Office, Cinemax, Showtime and the Movie Channel.

News: 24 hour per day news is offered on Cable News Network (CNN) based in Atlanta; a headline news service is offered on CNN 2.

Financial News: the Financial News Network provides stock market and futures quotations and advice from financial analysts.

Weather: national weather channels offer 24 hour per day forecasts; many cable systems add their own local weather channel.

Sports: all-sports channels cover everything from college football to hockey to wrestling, regional sports channels offer more coverage of regional teams and colleges.

Superstations: these are 'local' stations in cities such as Atlanta, New York, and Chicago that are now transmitted nationwide. They feature movies and syndicated programmes.

Music: The advent of music channels has spurred the growth of music videos, particularly for rock music; the major service is MTV (Music Television), although competitive channels are being introduced; the Nashville Network specialises in country and western music.

Children's Channels: high quality children's programmes on the Disney Channel, Nickelodeon and Discovery Channel.

Ethnic Programming: the Spanish International Network (SIN) offers Spanish language programming, mostly from Mexico, the major Hispanic markets nationwide; Galavision is a Spanish language pay television movie network.

Other ethnic channels include Black Entertainment Television (BET), and National Jewish Television (NJT).

Religious Channels: at least 6 channels offer sermons, spiritual music, religious education; these 'electronic churches' raise millions of dollars annually from televised solicitations. Also, the Mormon Church has an extensive closed circuit network and the Catholic Church runs a national educational network.

Educational Channels: the Learning Channel offers adult education courses for enrichment and credit; Lifetime offers health programming; other channels offer continuing education for professionals.

Specialised Entertainment: Examples include the Nostalgia Channel, Caribbean Super Station, regional entertainment networks, the Home Theater Network (for family viewing) and Daytime with specialised programming for women.

Adult Entertainment: 'restricted' channels such as the Playboy Channel and American Extasy (sic).

Congress: full time coverage of the House of Representatives is carried on C-Span; Senate coverage has also now begun.

Home Shopping: networks market directly to cable subscribers.

The economics of these new television services are complex. Some networks make their programming available to the cable operators free of charge. The religious broadcasters have turned satellite television into a major source of revenue through solicitation for donations from viewers. A second category includes programmers who offer their programmes free but who also carry paid advertising. Financial News Network is an example of such a channel that reaches more than 25 million subscribers. Pay television services such as Home Box Office charge a monthly fee per subscriber which is collected by the cable operator.

Pay-per-view charges for individual events such as boxing matches. On two-way systems equipped with addressable decoders, subscribers may order such services. However, these systems are not widespread, and most subscribers have to rent a decoder from the cable operator to receive such programmes, which are often also received by clubs and bars with their own TVROs.

Most channels charge the cable system operator a fee per month per subscriber. The cable operator recovers these charges through packaging the channels, generally as 'tiers'. The lowest tier carries local channels which must be carried according to FCC regulations (currently being challenged), local access placed on higher tiers for which there are additional monthly charges. A converter may be required to receive these signals. Movie channels are generally packaged separately with an additional monthly charge.

Satellite news gathering

Another recent satellite service is satellite news gathering (SNG), the use of portable satellite uplinks to cover news from the field. Many local stations are acquiring SNG trucks to enable them to cover breaking stories in their territory. This equipment is already having an impact on national news. Networks can now ask affiliates to cover a story and to transmit footage to be used in network newscasts.

The networks themselves have used this technology to air live coverage of international events such as the Philippine elections and 'people's revolution' and the Mexican earthquake. The flexibility of satellites in allowing transmission from any point, now with suitcase size units, will likely be increasingly significant as these SNG trucks suitcase-sized uplinks proliferate.

Teleconferencing

Another form of satellite broadcasting is video teleconferencing. Several types of conferencing are now common. Private teleconferencing systems link several branches of an organisation for regular meetings and consultations. These systems may be full duplex, with two way video (and associated audio and data) or one-way video with only audio/data return links. Small in-house networks may be fully two way, for example, to permit regular meetings between corporate headquarters and regional centres, or between engineers working on development of a single product. More common is a point-to-multipoint system.

225

Some corporations have adopted satellite teleconferencing as a regular in-house marketing tool. Merrill Lynch's marketing department transmits a weekly programme to branch offices around the country; the J.C. Penney department store chain presents new merchandise to buyers at its branches via satellite. Others use satellite facilities for *ad hoc* teleconferencing, i.e. single or infrequent video meetings. Trade associations may offer special seminars or short training courses to members around the country. Car manufacturers may introduce their new product line to car dealers. Unions may hold strategy sessions with members across the country.

Typically, these groups will rent the facilities they need for both transmission and reception; for example, they may rent transportable uplinks to transmit from a convention centre, or may use the facilities of a television station. Viewers can gather at locations with TVROs and meeting facilities. Several hotel chains now offer this service, many using the same equipment that receives pay TV for guests (primarily in the evening). Public broadcasting stations also rent out their facilities for groups that want to participate in teleconferences. Organisations or universities may also instal their own TVROs or rent them for special occasions.

Satellite television is also being used for training. Computerland retail stores have installed a network to train and update their sales staff and service personnel. An innovative use of satellite teleconferencing for education is the National Technological University (NTU). Established in 1985, NTU is an outgrowth of the Association for Media Based Continuing Engineering Education (AMCEE), whose members are universities that use television in engineering instruction. Many of these schools operate Instructional Television Fixed Service (ITFS) networks that transmit courses out to students in industries in the surrounding area. The companies pay to participate in the programme as well as paying for the tuition of their employees who are able to take graduate courses at work without having to travel to campus.

These schools videotape their courses and make them available as packages for group instruction through AMCEE, with credit being offered by the originating institution. NTU takes this concept several steps farther by offering courses via satellite direct to the work place anywhere in the country. Corporate members pay a fee to join the network plus tuition for each student. Students are able to interact with the instructor either during live classes through telephone links or by calling directly to consult with the instructor.

NTU has received accreditation and is just beginning to offer its first graduate engineering degrees.

In Texas, a satellite network called TIE-IN has been established to deliver courses to rural schools. A new curriculum requires Texas high schools to increase their offerings to include more instruction in foreign languages and advanced science and mathematics. Many school districts cannot afford or cannot find teachers with the necessary qualifications to teach the courses. The satellite network is being used to offer these courses where teachers are not available. The initial pilot project is rather small, but there are plans to expand the number of schools and courses offered if the system is determined to be cost effective.

Radio broadcasting via satellite

Satellites are also being used for radio broadcast transmission. Most programmers use digital transmission which results in very high quality transmission and better discrimination against adjacent satellite interference than using analog, thus making the use of smaller receiving antennas possible. Satellites are used not only by the national radio networks to feed news and special programmes to affiliates, but to create a new type of automated radio station. These stations typically offer one format of music (e.g. classical, hard rock, easy listening) with breaks for insertion of local advertising. The listener may not even realise that the programme does not come from a local studio.

In major cities, some broadcasters have found the satellite package a low cost means of competing with more established stations without having to hire popular announcers and to build extensive record libraries. In smaller cities and towns the satellite stations have introduced formats such as classical music and jazz that were not offered by local stations.

Voice and data services

Voice traffic (including voice circuits that have been conditioned to handle data traffic) is a major user of satellite capacity. The applications include light and heavy route trunks between switching centres and long haul circuits for private networks. Annoying echos that plagued early voice networks have largely been eliminated, although the delay caused by the distance of over 44,000 miles makes satellite voice transmission inferior to terrestrial means. However, in remote areas such as Alaska, satellite communication will continue to be the only viable means of long distance communication.

Satellites are also used for data transmission, ranging from very high speed transmission for high volume users to low-speed transmission of alpha numeric data, often on voice circuits conditioned for data transmission. As the number of VSATs and teleports increases (see below), data communication by satellite may grow.

VSATs and teleports

Another new satellite technology is Very Small Aperture Terminals (VSATs), also known as micro earth stations. The receive-only versions are generally less than one metre in diameter and are used for data broadcasting, i.e. point-to-multipoint data reception. There are now more than 20,000 receive-only VSATs installed in the U.S., typically replacing leased lines because of their lower cost and higher reliability. VSATs are used by wire services to reach newspapers nationwide, by the National Weather Service to transmit weather information, and by brokerage and financial services to transmit financial data to brokers and commodities traders.

More recently, interactive VSATs have been introduced for data communications such as links from field offices to headquarters, and client transactions such as car rentals, credit card verifications and automatic teller machines. These are all examples of 'bypass' using satellites instead of telephone lines. VSAT technology is attractive because the carrier can guarantee high quality and fixed user costs, whereas with deregulation, terrestrial networks are becoming more complex, with several telephone companies involved; local loop charges are increasing; and significant delays are encountered in installing and modifying facilities.

Teleports are communications hubs designed to provide a variety of services including voice, data, and video through both domestic and international satellite links. A teleport typically consists of several antennas pointed at different satellites with microwave or fibre optic links to users. Customers include banks, teleconferencing services, and broadcasters.

The New York teleport on Staten Island is a joint venture of Merrill Lynch and Western Union. In addition to antennas for communication with all domestic and Atlantic Ocean satellites, the New York teleport provides dedicated fibre optic links to its customers. In Washington D.C., antennas on the National Press Building transmit television news feeds to a teleport outside the city. In Houston, the teleport also serves as a hub for VSATs and is located on customer's buildings, to provide communications among company branches around the country. Similar teleports are being built in many U.S. cities.

Direct broadcasting satellites

The commercial viability of direct broadcasting satellites (DBS) that would bring multiple channels of television to every household via very small rooftop antennas, as predicted a decade ago, is now highly questionable. While more than one million individuals have installed backyard antennas, and antennas to receive special signals are sprouting on many buildings, it appears unlikely that DBS will reach the penetration of cable, since most of the country is already cabled.

The current high powered satellite technology cannot offer as many channels as are available on most cable systems. In addition, DBS has not been able to show a price advantage over cable. Initially, DBS operators proposed selling antennas that would allow owners to pick up future programming at no monthly charge. However, low projections of penetration and lack of sufficient advertiser support soon led DBS developers to realise that they would have to charge a monthly fee. Only if DBS operators can find a means to significantly undercut cable fees or to offer alternative programming so attractive that it lures viewers from cable, will urban DBS take off.

Although originally DBS programmers planned to offer programming not available from other sources, there does not seem to be a niche that has not been filled by another medium including broadcast TV, cable TV, pay TV, and VCRs. One industry analyst states that the financial incentive that will make DBS succeed is the additional revenue that can be generated for programme suppliers by delivering their programme to households without cable.[5] Higher powered Ku-band satellites and addressable descramblers will make it possible for satellite broadcasting to reach non-cabled households with the same variety of programmes and comparable rates to cable systems. The four parties of interest, the programme producers, the programme suppliers, the satellite operators, and the cable systems, will endeavour to find means to use DBS to increase their revenues and protect their interests. Cable companies for example, may market DBS services for low density areas surrounding their franchise area where cabling costs are prohibitive.

While it appears that DBS is attractive for non-cabled rural areas, many of these viewers have already installed C-band antennas to pick up the channels being fed to cable systems. Even if more of these cable channels are scrambled, it is still likely to be more attractive to offer the same channels for direct reception by selling or leasing descramblers than to offer a separate satellite network for direct reception.

However, cable programme providers would clearly prefer to deal only with cable operators rather than with individuals. One concern the programme providers have raised is the complexity of billing individuals. A strategy to eliminate individual billing is being developed by the Rural Electric Co-operatives which are proposing to become brokers for a package of satellite TV channels. They would negotiate with the distributors on behalf of their customers for a package of satellite channels. They would then collect monthly fees using the billing system already in place for rural electrical services, thus eliminating billing requirements for the distributors. They would also sell or rent satellite antennas to their customers.

Several bills have been introduced into Congress to protect the interests of backyard antenna owners. For example 'The Rural Satellite Dish Owners Protection Act' seeks to prohibit the encoding of satellite transmitted television programmes until decoding devices are fully available at reasonable prices (S. 2290). 'The Satellite Viewers Rights Fair Disclosure Act' of 1985 would require any person who sells earth stations for private satellite reception to display a public notice stating that some channels may be scrambled, that descramblers are available, that earth station owners may be required to pay for channels they receive, and that unauthorised reception of network programming and other programming not intended for cable television is strictly prohibited. (HR 4414)

Other bills seek to delay or prohibit scrambling of satellite signals. While none of these bills was passed during the 99th

Congress, they show the strength of the predominantly rural lobby to preserve direct access to satellite television.

FUTURE OF SATELLITES

The explosion of the Challenger, the failure of a French Ariane launch, and the failure of a Delta rocket launched from Vandenberg Airforce base in California, all in the first five months of 1986, temporarily halted the launching of geostationary satellites by the western world (whose customers include developing countries such as India, Indonesia, Brazil, Mexico as well as the OECD nations). However, this catastrophe will eventually be considered only a temporary setback, although one which may have a permanent impact on the satellite insurance and the mix of rocket versus reusable (shuttle) launchers. At present, there is an oversupply of transponder capacity, resulting in a drop in leasing charges and a 'buyers' market'. However, the delays in new launches as a result of the 1986 failures may delay deployment of additional capacity.

Another factor of greater long term significance is the installation of optical fibre on high volume routes, both intercity and international. For example, the next transatlantic cable will be optical fibre. It is unclear at this point to what extent optical fibre will divert existing and projected traffic from satellites. The deregulation of the telecommunications industry has resulted in several carriers building their own long distance optical fibre networks. Fibre optics have an advantage on fixed route point to point high density networks where the traffic volume justifies the enormous bandwidth. The low cost per circuit mile of fibre is dependent on high traffic volume to cover fixed costs. As Inglis points out, the low cost per circuit mile is comparable to claiming that the Boeing 747 has the lowest cost per mile of all commercial aircraft. This is true, but it is only significant if most of the seats are filled with revenue passengers.[6]

Fibre optic costs per circuit mile are high for low utilisation but decline rapidly as the number of circuits in use increases; satellite costs per circuit are lower at low utilisation but decline more slowly. While fibre will take over the heavy volume point to point traffic, it may be complementary to satellites for other applications such as local area networks that will feed data to office building rooftops, and short haul links to connect users to teleports.

In the broadcasting field, satellites will likely continue to have major advantages. For example, the point-to-multipoint distribution capability of satellites is unmatched by any terrestrial alternative. Thus it is possible for a broadcaster to transmit to all of its affiliates across the country. Transmissions may originate from any point, as both fixed and transportable antennas may be used for uplinking. Similarly, pay TV operators can feed cable systems across the country.

In the future, business customers may be the greatest users of satellites for specialised data communication networks and for 'business television'. Voice telephone service will migrate to optical fibre except in remote areas where satellites remain the only viable solution. And satellite television will fill in the gaps by reaching those households directly that do not have cable, rather than replacing urban cable systems. Cable is a higher capacity and more flexible medium. Direct satellite reception will be used only in areas where cable distribution is uneconomical, approximately 15 to 20 million households (or about 20% of U.S. television households).

It appears that satellites will continue to play a major role in point-to-multipoint services such as broadcasting, distribution of services to cable systems, broadcast data networks (weather, stock prices, etc.) and video teleconferencing. Satellites also appear to be the least cost solution for a variety of new multipoint-to-point interactive services such as credit card verification, field office to home office communications, and remote data collection and monitoring. Thus, it is safe to say that communication satellites will remain an important element in U.S. telecommunications of the foreseeable future, although their role may differ from what was predicted twenty years ago and even five years ago.

REFERENCES

1. Hudson, H. and Parker, E.B., 'Medical Communication in Alaska by Satellite', *New England Journal of Medicine*, Dec. 7, 1973.

2. Foote, D., Hudson, H. and Parker, E.B., *Telemedicine in Alaska : The ATS-6 Biomedical Demonstration*, Institute for Communication Research, Stanford, 1976.

3. Oslund, J., "Open Shores' to 'Open Skies' : Sources and Directions of U.S. Satellite Policy', in Pelton J. and Snow, M., *Economic and Policy Problems in Satellite Communications*, 1977.

4. Quoted in Oslund, J., "Open Shores' to 'Open Skies'", 1976, p. 145.

5. Inglis, A.F., 'The United States Satellite Industry — An Overview', *Proceedings of FIBRESAT 86*, Vancouver, Canada, Sept. 1986.

6. Inglis, A.F., 'The United States Satellite Industry — An Overview', 1986.

Other useful sources of information:

Baylin, F. and Brent, G., *Satellites Today: The Guide to Satellite Television*, Howard W. Sams and Co., Columbus, Ohio, 1986.

Channel Guide, vol. 6, no. 4, November 17-23, 1986.

Cook, R. 'Satellite Communications at Down-to-Earth Prices.' *High Technology*, December 1986.

Demmert, J.P. and Wilke, J.L. 'The LEARN/ALASKA Networks: Instructional Telecommunications in Alaska' in *Telecommunication in Alaska*, (ed) R. M. Walp, Pacific Telecommunications Council, Honolulu, 1982.

Filep, R., 'The World Communications Satellite Market Through 2000: Analysis and Forecast.' *Satellite Communications: Developments, Applications, and Future Prospects*. Online Publications, London, 1984.

Gavin, J. Jr. 'The United States Industry'. in Kirton, J. (ed), *Canada, The United States, and Space*. Canadian Institute for International Affairs, Toronto, 1986.

Glatzer, Hal., *The Birds of Babel*. Howard A. Sams and Co., Indianapolis, 1983.

Hudson, H.E., 'Medical Communication in Rural Alaska'. in *Telecommunication in Alaska*, (ed) R. M. Walp. Pacific Telecommunications Council, Honolulu, 1982.

'New Ways to Keep in Touch.' *U.S. News and World Report*, April 28, 1986.

Parker, W.B., 'The Evolution of the Present Alaska Telecommunications System'. in *Telecommunications in Alaska* , (ed) R. M. Walp. Pacific Telecommunications Council, Honolulu, 1982.

Smith, I. V., 'Retail Gains'. *Satellite Communications*, April 1986.

Smith, I. V., 'Taking the Bull by the Horns', *Satellite Communications*, May 1985.

Sparks, W., 'Satellites, The Telegraph, and the Pony Express'. *Satellite Communications: Developments, Applications and Future Prospects*. Online Publications, London, 1984.

Williamson, D. Jr., 'Changes and Choices in United States Space Policy'. in Kirton, J. (ed) *Canada, The United States and Space*. Canadian Institute for International Affairs, Toronto, 1986.

Wines, L., 'Business Tunes in on Telecommunications'. *Crain's New York Business*, July 8, 1985.

9

Satellite Communications in Canada

Heather E. Hudson

THE BEGINNING

Canada became one of the earliest members of the global satellite community with the launching of its first satellite in 1962. As in the U.S., the major impetus for space communication research in Canada came from the military. The first Canadian satellite, Alouette, was funded and designed by the Defence Research Telecommunications Establishment and launched for Canada by NASA. Interestingly, its purpose was not to test satellite communications, but to collect data on the ionosphere from above. The propagation effects of the ionosphere were important for high frequency radio communications used by the military and also at that time for civilian communications in the remote north. Alouette I was followed by two other experimental satellites, Isis I and II, both designed and built in Canada and launched by NASA, to continue research on the ionosphere.

THE ANIK SYSTEM

Canada became the first country to use a geostationary satellite for domestic communications with the launching of Anik A-1 in 1972, and the back-up, Anik A-2, in 1973. (The U.S.S.R. was already using satellites in nongeosynchronous polar orbits for domestic telecommunications.) The name Anik, the winning entry in a contest sponsored by Telesat, means brother in Inuktitut, the language of the Inuit or Eskimos. It was chosen to symbolise the commitment of the government to use satellites to improve communications in the far North, but years passed before that promise was fulfilled. There appeared to be two other major reasons behind Canada's decision to establish a communication satellite system as proposed in 1967 in a cabinet White Paper. The first was to stimulate the development of a Canadian space industry. The second was to secure optimal orbital locations before the U.S. claimed them. At the time, the U.S. had put a freeze on commercialisation of satellite communications until it could be determined how the industry would be structured: i.e. publicly or privately owned; monopolistic or competitive (see chapter on U.S. satellites).

The early days of satellite service in Canada were noteworthy primarily for the absence of use of the system. As was the case with many new technologies, institutional rather than technological barriers stood in the way. The telephone companies had no incentive to expand service into the remote north because of the low traffic and projected high costs. Where required by regulators to provide or extend service, they preferred to use terrestrial technologies such as microwave because they could own the equipment and put it in the rate base. Satellite transponders and earth stations were leased from Telesat Canada, and thus could not go in the rate base, although the actual costs could be passed through to the customer. And although the cost of using satellites is independent of distance, the tariffs continued to be based on point-to-point distance, despite pressure for flat rates for intra-north calling.[1]

The only broadcaster to use the satellite was the CBC, which was required under the Broadcasting Act to provide service to all Canadians 'as funds became available'. The private broadcasters, having no such mandate and finding markets small and satellite transponder lease charges high, stayed away from Anik for nearly a decade.

NORTHERN BROADCASTING

Before Anik, the CBC relied on several rudimentary means of reaching the North. Radio, in the form of the CBC Northern Service, was produced in Montreal and transmitted via shortwave from transmitters in New Brunswick designed for international broadcasting. The signal was variable according to season, time of day, and ionospheric conditions, and generally reached only northern Quebec and the eastern Arctic.[2] Five local CBC radio stations in the north produced programmes for their immediate area (in the Mackenzie region and the Yukon they were redistributed by terrestrial links). Vast areas of the North, especially the central Arctic and the northern parts of the provinces had no CBC radio service (nor did they have service from any other source). CBC television distribution was even more limited. The five northern communities with CBC radio plus a few other towns received the Frontier Coverage Package of CBC videotapes played over local low powered transmitters.

With the advent of Anik came the first live television transmissions to the far North. CBC network programming, most of which originated in Toronto, was transmitted to the Arctic on two transponders in Atlantic and Pacific time (the Arctic actually stretches across five time zones). Regional news from St. John's, Newfoundland, or Vancouver was inserted, but there was no northern oriented or native language programming.

Television quickly became popular among native and non native northerners alike. Indeed, the primary beneficiary was likely to be the Hudson's Bay Company (the modern version of the fur trading post) with general stores throughout the North. Early Hudson Bay ads proclaimed: 'Magic from the Sky! Television from Anik! Get your TV set at the Bay'.

As early as 1970 (at a conference in Yellowknife, the capital of the Northwest Territories, native leaders expressed concern about the likely impacts of 'southern' television and stated that there must be content about the North and in native languages. A decade passed before the CBC took the first steps to include indigenous programming.

The record for radio, however, was quite different. From the earliest days, the CBC Northern Service used the satellite to feed programming including native language content to isolated communities. The difference was that CBC radio was already established in the North, with five radio stations across the Arctic. These had been local stations that produced much of their own content, and re-broadcast network and some other programmes received by shortwave before the satellite era. With the advent of Anik, the stations were able to receive the CBC network to add to their own local production. But the local stations became regional network hubs and redistributed their signals out to the villages in their region.[3] Thus they received a mixture of national and regional content with a significant native language component. Under an innovative CBC community radio project, each local community could contract with the CBC to produce a limited amount of local programming which could be substituted for network programmes. In many cases the CBC trained local people and provided a small studio placed in a building supplied by the community. Most stations were run by volunteers with operating expenses covered with locally raised funds.

Northern television took much longer to evolve, despite continued protests from native leaders about the negative cultural and linguistic impact of CBC network television. The opportunity to demonstrate an alternative approach came with the experimental Hermes satellite, followed closely by Anik B.

EXPERIMENTAL SATELLITES: CTS AND ANIK-B

The Communications Technology Satellite (CTS) or Hermes was a joint project of the Canadian Department of Communications and NASA. Its high powered Ku-band transponder provided the first opportunity to experiment with direct broadcasting to very small antennas (less than one metre in diameter).

CTS was followed by Anik-B, a dual frequency C- and Ku-band satellite. The Department of Communications leased much of the satellite capacity to continue experimentation and demonstrations begun on Hermes. The third generation of Anik satellites, Anik-C is a Ku-band system, whereas the fourth generation, Anik-D is again C-band, replacing the now ageing Anik A. The first three systems had been built in the U.S. (Anik-A and C by Hughes Aircraft and Anik-B by RCA) with subcontracts to Canadian firms. The prime contract for Anik D, however, was awarded to Spar Aerospace, a Canadian firm nurtured by the earlier subcontracts.

At the time of its launch in 1976, CTS was the most powerful satellite built to date. It lasted four years, during which time it was used for experiments and demonstrations in telemedicine, distance education, and direct broadcasting to northern rural and remote communities. In retrospect, viewed as a 'concept tester'[4], CTS was followed by Anik B, which the Department of Communications viewed as a vehicle for pre-operational trials to determine the viability of new services.

SATELLITE EXPERIMENTS

During the Hermes period, from 1976 to 1979, 37 experiments were conducted in distance education, telemedicine, teleconferencing, and community communications and broadcasting. During the Anik-B programme from 1978 to 1984, 32 pilot projects were conducted of which 12 involved tele-education.[5] A few of these have made the transition to operational services on a commercial Anik satellite.

One of the successful experiments involved TV Ontario (TVO), officially the Ontario Educational Communications Authority. TVO used the Anik-B satellite to deliver educational programmes previously available only in southern and eastern Ontario to 46 underserved communities, directly to schools, libraries, cable TV systems, master antenna systems, and low power TV transmitters. TVO has now provided province-wide coverage by satellite using Anik-C to deliver its programmes throughout Ontario, and to replace the costly microwave links previously used to link major centres.

Another application of the TVO network has been to deliver native language radio programming throughout remote northern Ontario. A pilot project on Hermes demonstrated the production and delivery of native language radio programming from a production centre in Sioux Lookout, Ontario to remote Indian communities with their own community radio stations that re-broadcast the programmes to their communities and originated programmes themselves on a rotating basis. Listeners were very enthusiastic about the regional programming, and the sponsoring organisation began several years of effort to find an affordable means to continue and expand it. However, the tariffs set by Telesat and the telephone company were far beyond the means of the organisation. Eventually an agreement was reached with TVO that enables Wawatay to piggyback its audio signal on a TVO transponder. At the community, the signal is split, with the TV component being re-broadcast over a low power transmitter and the radio signal delivered to the community radio station where it is retransmitted.

In British Columbia, educators began an experiment in distance education on Hermes which was further piloted on Anik-B and is now an ongoing service on Anik-C. The Knowledge Network delivers credit educational programmes to isolated British Columbia Communities where they are retransmitted over cable systems or viewed at community centres and colleges. An audio conferencing link enables students to interact with their instructors based in the Vancouver/Victoria area.

Inuit have also used the experimental era to demonstrate and implement new broadcasting services for the North. Inuit Tapirisat of Canada, the national Eskimo organisation, carried out a pilot project on Anik-B which linked communities and produced programmes.

Another experiment on CTS linked eight Inuit communities in Arctic Quebec with an audio conferencing network. The network was used to broadcast two hours of Inuktitut language programming every other night for two months in 1978. All of the communities (none of which at that time had television service) could participate through the conferencing link.

ITC's Inukshuk project (named after stone cairns that serve as guideposts in the Arctic) linked six Inuit communities in 3 Arctic regions (with different dialects and time) in a one-way video, two-way audio conferencing network. The video signal was viewed in each community on a large screen in a meeting room and in five communities it could be re-broadcast to reach people at home. Interaction from each site was facilitated with co-ordinators. The project produced more than 320 hours of programming during a nine month period of which approximately 50% was live, generally interactive teleconferences.[6] The project not only fostered communication among Inuit across 3 regions about issues of land claims, education and cultural identity but demonstrated to all including the federal government, the Northwest Territories government and the CBC that the Inuit were capable of producing their own programmes.

The Inuit established the Inuit Broadcasting Corporation in 1981; it now broadcasts five hours of Inuktitut programming each week using time on the CBC Northern Television Service channels. Uplinks are located in Frobisher Bay (on Baffin Island) and in Salluit in Arctic Quebec.

In 1983, the federal government recognised the importance of native broadcasting in its Northern Broadcasting Policy which established the Northern Native Broadcast Access Programme (NNBAP). NNBAP is providing $40.3 million over four years to thirteen northern native communications societies to enable them to establish production facilities, train broadcasters, and produce and distribute native language and native oriented programming. Many of the organisations distribute their programmes via satellite.

239

CABLE TELEVISION

Some 98.1 percent of all Canadian households have at least one television set, with 87% of television households owning at least one colour set. Satellite broadcasting in Canada is intertwined with cable television. Canadians were early adapters of cable technology as a means of receiving U.S. channels. By 1961, there were already 200,000 subscribers and 260 cable systems. By 1982, there were 564 operating cable systems with about 5 million subscribers. About 80% of Canadian homes passed by cable subscribed to the service. Penetration rates in some cities are extremely high: for example, in Toronto, 77% of households are subscribers, while in Vancouver, 91% of households are cable subscribers.[7]

Canadian cable systems are federally licensed and regulated by the Canadian Radio-Television and Telecommunications Commission (CRTC) which sets policies on ownership, fees, and content. (For a discussion of the history of cable regulation in Canada see Caron and Taylor, 1985.) Canadian cable systems must carry local CBC and other local educational and commercial channels, regional CBC (unless it duplicates local CBC programming), all other regional programming, extra-regional CBC, extra-regional educational channels, community programming, and any other extra-regional stations that do not duplicate higher priority stations. Once these priorities are met, they may carry foreign stations according to a 3 plus 1 import limit of three commercial signals and one non-commercial signal. Thus, until recently, Canadian cable subscribers received only the four U.S. networks, generally picked up from U.S. over-the-air broadcasts at the border and retransmitted by microwave to the cable system.

The first satellite-delivered special programme available to Canadian cable subscribers was the House of Commons, transmitted via Anik and now carried live on cable systems across Canada. While the Commons debates are not tops in the ratings, they do give citizens and students in particular an opportunity to see their representatives in action, and they provide footage which is often inserted in network and local newscasts. Another well established service is a package of programmes received from the French TDF satellite and retransmitted to cable systems in Quebec by a consortium of cable operators called 'La Sette'.

PAY TELEVISION

By the early 1980s a policy anomaly had developed in that owners of individual backyard and community earth stations were receiving (and in some cases redistributing) U.S. satellite channels, whereas the cabled urban areas received only the U.S. networks because CRTC regulations prohibited cable operators from carrying these signals. Faced with the threat of U.S. pay TV, the CRTC decided to license Canadian pay TV operators, although the federal Department of Communications had not yet formulated a national pay television policy.

In 1980, the CRTC adopted a policy of authorising Canadian pay television channels as a means of countering the threat of U.S. satellite delivered programming. The first of these channels began operation in 1983. The CRTC chose a form of 'controlled competition', authorising a limited number of Canadian owned pay TV channels. The policy initially amounted to U.S. content being offered via a Canadian satellite, as most of the movies, sports and music are from the U.S.

In order to prevent excessive market fragmentation, the CRTC limited the number of pay services in any region to four. The licences were awarded to First Choice Canadian Communications to provide 24 hour movie and entertainment channels in English and French; Lively Arts Market Builders for a national cultural channel ('C-Channel') of performing arts, films, quality children's programmes, etc.; five regional film and entertainment channels, and one regional multi-lingual channel.

The CRTC placed conditions on the licence in order to ensure the delivery of Canadian content. For example, First Choice was required to devote 60% of its programming budget to Canadian material and 50% of that to drama. In order to protect Canadian broadcasters, C-Channel was not allowed to fill more than 40% of its schedule with films, and not more than 5% to films that had been among the top thirty grossing films in Canada in the past three years.

Operating under these constraints, pay television has floundered. C-Channel folded five months after it began operation. (Two cultural channels in the U.S. were also cancelled in 1982-83. Promoters stated that the market was much smaller than they predicted; perhaps more significantly, a longer term commitment was needed to test the viability of national cultural networks.) The specialised entertainment channels have also suffered bankruptcies and mergers. There is now one major English language and one French language movie channel and a music channel. Cable operators are now allowed also to offer U.S. cable channels (such as CNN news, financial services, sports) on higher tiers.

DBS VERSUS CABLE

In 1983, the CRTC also licensed advertiser supported Canadian cable channels. But even that strategy was thought insufficient to counter the popularity of U.S. satellite TV. Canadian broadcasters have apparently concluded that their future is tied to cable. Under Canadian 'must carry' rules (described above) their signals are included on the basic tier of cable systems; however, if Canadians find direct reception from U.S. satellites more attractive than cable, they may not even be exposed to Canadian programmes. Thus, even broadcasters have endorsed a CRTC decision to allow carriage of up to five U.S. advertiser supported cable channels. Now urban Canadians can see what their rural counterparts have been watching for several years.

Thus it appears that satellite broadcasting in Canada is to be limited to distribution to cable systems (and low power transmitters in small communities) despite the attention given to direct household reception a decade ago. The investment in cable in Canada has made it the standard delivery system for multiple channels. High powered direct broadcast satellites could not deliver as many channels and would require replacing the cable infrastructure with individually owned antennas. Medium power satellites can be used to feed cable head ends, and to provide direct reception in rural areas such as to farms and ranches where population density is too low to justify community redistribution.

POLICY ISSUES

The challenge of new technologies

New technologies challenge the existing framework of regulation and policy making. For example, in Canada, individuals and communities installed satellite antennas to pick up signals from U.S. satellites. To them this was a broadcasting service, with the signals literally there for the taking (although some groups offered to pay the U.S. originators). To the federal Department of Communications this was piracy, because individuals were receiving signals from fixed service satellites that were intended only for redistribution and not direct reception. The courts did not resolve the issue, and the government was forced to turn a blind eye on the politically vociferous (and primarily rural and isolated) antenna owners.

The government could keep most urban Canadians from watching U.S. satellite channels by refusing to allow licensed Canadian cable systems to carry them. However, individual reception and the growth of VCRs were already undermining the Canadian content policies. Thus the CRTC acted to introduce Canadian pay television before the government's policy arm (the Department of Communications) was ready with a pay TV policy. This was one in a series of cases in which the policy makers accused the regulators of pre-empting their function. However, the regulators felt obliged to tackle the world as they found it and to respond to public and industry pressure to authorise new services.

Slow growth of educational services

The growth of educational applications of satellite broadcasting has been slower than an observer might expect, given the number of small and isolated communities that could benefit from distance education. The major reason is that education is a provincial responsibility in Canada, so that although the satellite can cover the country, there is no standardisation in educational content or accreditation at the national level. The Knowledge Network in British Columbia has developed a successful model, but it has not been emulated or extended to other provinces. TV Ontario's programmes are now available throughout Ontario and can be received elsewhere, although they are used more for enrichment than formal instruction. There is as yet no national educational channel or university channel as is found in the U.S. (see chapter on U.S. satellite broadcasting).

Native broadcasting

Perhaps the greatest success in the Canadian satellite experience to date has been in native broadcasting. It took more than a decade for native people in the north to obtain the benefits promised by the government in the late 1960s. However, during that period a foundation of community radio and a development-oriented approach to media was built. One strong influence was the National Film Board's Challenge for Change programme which used film as a participatory community development tool. The Department of Communications funded a pilot project in northern communications in the early 1970s with a community development approach. The French concept of 'animateur' or animator was applied to media projects from urban video and cable access to remote community radio. The federal Department of Communications and the CBC supported community media projects while the CBC supported community radio through facilitating access to CBC radio transmitters. Pilot projects with federal and provincial support explored a range of community media models from local access to TV transmitters to portable video to community radio. Thus, when native people began to gain access to satellites (at first through experiments and pilot projects on Hermes and Anik B), the community-oriented developmental model was already well in place. This decentralised, participatory and developmental use of satellites is being adopted by aborigines in Australia and may be a useful model for other developing regions.

Satellite policy

Telesat Canada was established by the Telesat Canada Act in 1969 to own and operate Canada's communication satellites. The act stipulates that the corporation is to be owned by the carriers, the government and the public. However, to date, public shares have not been offered. Hence, at present, Telesat is owned 50% by the government and 50% by the carriers (of which Bell Canada is the largest shareholder with a 24.6% interest). Telesat acts as a 'carrier's carrier', and thus does not deal directly with the public, except that it can contract directly with broadcasters to supply transponder capacity and facilities.[8]

As a result of its ownership and structure, Telesat has never been able to compete aggressively with the established carriers as satellites have in the U.S., but acts as a supplement to their terrestrial facilities. Telesat has not been as innovative in its service offerings or pricing as U.S. satellite systems.

In 1975, Telesat decided that it should become a member of the Trans-Canada Telephone System (TCTS), now Telecom Canada, which is a consortium of Canadian telecommunications carriers, to facilitate integration of facilities and obtain additional financial resources for planning and construction of its next generations of satellites. In early 1976, the federal government approved the plan, subject to the approval of the CRTC. However, after extensive hearings, the CRTC rejected the plan, believing that an independent Telesat might better serve the public interest in fostering a competition in long haul data, video, and other private line services (Telecom Decision 77-10). Under intense lobbying from the carriers, the Governor-in-Council, however, varied (i.e. reversed) the decision, stating that the public interest would be served better by providing Telesat with access to the carriers' financial resources and encouraging the carriers to use the system. Thus unlike in the U.S. where an FCC decision would be appealed through the courts, in Canada a 'political appeal' of a regulatory decision may be made directly to cabinet.[9]

The CRTC once more attempted to foster competition in 1981 by ordering Telesat to offer service to end users and to lease less than full transponders in order to meet the needs of smaller carriers and end users. However, once again the Governor-in-Council varied the decision to maintain Telesat's carrier's carrier role, except to confirm that Telesat be allowed to deal directly with broadcasters (Order in Council P.C. 1981-3456, Dec. 8, 1981). By 1985, Telesat was proposing to amend its definition of 'customer' so that it could deal directly with end users, but it remained firmly entrenched in the Telecom Canada family.

Earth station ownership

While originally, only Telesat could own satellite earth stations which it then leased to customers, this policy has gradually changed in recent years. In 1979, broadcasters and common carriers became able to license receive-only earth stations, and common carriers could receive licences for their own 14/12 GHz transmit stations.

Not until 1983 could individuals and small commercial establishments own their own receive-only stations without a licence (long after thousands of Canadians had purchased TVROs to pick up U.S. satellite television). In 1984, DOC announced a two-step process to liberalise uplink ownership for broadcasters and business users (DOC NR-84-5265E, April 10, 1984). The result has been a dramatic reduction in the cost of earth stations for users, and a decline in equipment leasing revenues for Telesat.

Broadcasters now have the right to purchase services directly from Telesat rather than ordering them through a common carrier, and they may now resell excess transponder capacity to other broadcasters.

Culture versus choice

In 1982, nearly 75% of all English language viewing time by Canadians was spent watching U.S. programmes (primarily entertainment and drama) on U.S. and Canadian channels. Some 96% of entertainment drama — films, soap operas, detective series, plays etc., — was foreign produced, from the U.S. and U.K. The Canadian policy dilemma was summarised in 1983 by then Minister of Communications Francis Fox, who stated that '... within a healthy and viable Canadian broadcasting system Canadians are entitled to as much choice in programming as possible'. He added: 'I also firmly believe that 'choice' for Canadians is meaningless unless it also includes programming which reinforces the cultural heritage of all Canadians.[10] Yet Canadians do watch domestic news, sports, current affairs and documentaries and this behaviour, which was first studied by the author in 1969[11] has continued with the availability of more foreign programme choices through pay TV.

IMPLICATIONS FOR OTHER COUNTRIES

The Canadian case may be illustrative for European policy makers in the process of authorising new satellite and cable services. Canada's Federal Cultural Policy Review Committee stated: '... if Canada is to retain a programming presence in its own broadcasting and telecommunications system, it must use satellites technological and creative resources to provide Canadian programmes and services that Canadians want to see and hear, programmes that are competitive in quality with those from other countries'.[12] However, U.S. entertainment programming has remained overwhelmingly popular. Thus, rather than declaring viewing of U.S. satellite delivered channels illegal, Canadian policy makers have sought to use them to keep subscribers happy, while encouraging the delivery of as much Canadian produced programming as they can find means to support.

Canada's dilemmas concerning satellite broadcasting may soon be shared by other nations as satellites and new video technologies proliferate. The channel capacity of cable and satellite delivery systems is enormous. National producers may find that they lack the resources to produce sufficient content to fill the channels or to attract an audience to pay for the production and delivery systems. The demise of several Canadian pay channels and the demonstrated preference of Canadians for American entertainment programming either on cable, backyard satellite antennas, or video cassettes may indicate future trends in other countries attempting to both introduce new technologies and foster their own cultures and production industries.

REFERENCES

1. Inuit Tapirisat of Canada, 1978.
2. Hudson, H.E., 'The Role of Radio in the Canadian North', *Journal of Communication*, Autumn 1977.
3. Hudson, 'The Role of Radio in the Canadian North', 1977.
4. In Heather Hudson, (ed) *New Directions in Satellite Communications : Challenges for North and South*, Artech House, Dedham, MA, 1985.
5. Davies, N.G. 'Canadian Space Applications: Models for the Developing World', 1985.
6. Valaskakis, R., Robbins, R. and Wilson, T. 'The Inukshuk ANIK B Project : An Assessment', Inuit Tapirisat of Canada, Ottawa, 1981.
7. Caron, A.H. and Taylor, J.R., 'Cable at the Crossroads: An Analysis of the Canadian Cable Industry' in Negrine, R. (ed) *Cable Television and the Future of Broadcasting*, Croom Helm, London, 1985.
8. Bruce, R.R., Jeffrey, P.C. and Mark, D. (ed), *From Telecommunications to Electronic Services*, Butterworths, London, 1986.
9. Kaiser, G.E. 'Developments in Canadian Telecommunications Regulation' in Marcellus Snow (ed), *Marketplace for Telecommunications* Longman, New York, 1986.
10. Quoted in Hollins, T., *Beyond Broadcasting: Into the Cable Age*, British Film Institute, 1984.

11. Hudson, H.E., Unpublished Paper, 1969.

12. Quoted in Hollins, *Beyond Broadcasting*, 1984.

Other Useful Sources of Information:

Kirton, J. (ed), *Canada, the United States, and Space*, Canadian Institute for International Affairs, Toronto, 1986.

See also the following in Kirton, J. (ed), 1986:

Collin, A., 'The Canadian Space Program'.

Curran, A., 'The State of Canada's Industry'.

Golden, D., 'The Uncertain Future of Domestic Commercial Communication Satellites'.

Murphy, B., 'Satellites in Canada: Past, Present and Future', *Proceedings of FIBRESAT 86*, Vancouver, Canada, 1986.

Valaskakis, G. 'Socio-Economic Implications in Canada of Satellite Communications', *Proceedings of FIBRESAT 86*, Vancouver, Canada, 1986.

10

Direct Broadcasting by Satellite in Japan : an Overview

Yuko Nakamura with an introduction by Ralph Negrine

A Brief Introduction to Broadcasting in Japan.
By Ralph Negrine

Broadcasting in Japan is controlled by two types of organisations. NHK, the publicly funded public service broadcasting organisation, is by far the most important of these two types. It is funded through a licence fee system and is required to present high quality programmes that satisfy the demands of the public and also elevate the cultural level of the country. It runs two television services, the General Channel and the Educational Channel, which are available to nearly all of Japan's 38 million households.

The other type of organisation which is concerned with Japanese broadcasting is the commercial/private organisation. Originally, individual private companies were licensed to cover specific 'prefectures' within the Japanese territory, but commercial pressures have led to the amalgamation of many of these companies and to the creation of 5 television networks based in or around Tokyo.

One important effect of this private/public mix is that it does offer a choice of programming. The ordinary television viewer in Japan has a wide range of material to choose from and this increases the difficulties of entrepreneurs wishing to exploit the new media by introducing new services which are, in effect, no different from what already exists. In a review of Japan's new media, Kobayashi argued that the new media face difficult problems. He observed that in Japan 'a multi-channel trend seems less strongly motivated, because a certain level of diversification of broadcasting services has been already attained and future broadcasting policies are still uncertain'.[1]

Equally important is his belief that there is no new significant source of programming material for the new media. The situation is an extremely competitive one and the material currently produced within Japan just about serves the existing channels. Broadcasters—and supporters of new media channels—are not always able to turn to overseas producers since much of that material, including material beamed by satellite to Japan (e.g. CNN), is already incorporated in the schedules of the terrestrial broadcasting systems.

This difficulty with programming may account for the failure of cable television to take off. It still covers only 13% of Japan's households and it does not look as if its fortunes will improve dramatically in the immediate future despite governmental commitments.[2]

In this context, and bearing in mind that the Japanese population appears well satisfied with the mix in its broadcasting structures, the policies towards direct broadcasting by satellite may seem difficult to comprehend. To quote Kobayashi again 'There is widespread controversy on the merits of high power DBS, compared with low-power DBS and the advance of cable TV linked with communication satellites. For the time being in Japan, however, there are no plans to reconsider the original DBS plans..'.[3]

The Ministry of Posts and Telecommunications commitment to DBS remains unchanged and this also applies to NHK. Together, for example, they funded the current DBS service, BS-2 and they are presently considering BS-3 as the chapter below indicates. But these commitments continue to be fairly controversial. Despite the Ministry's desire to see Japan leading the world in satellite technology and NHK's desire to eliminate the remaining areas of bad reception, there are still reservations about the wisdom of DBS. For example, NHK intends to use the DBS facility to reach the outlying islands of Japan and a mere 400,000 homes without any terrestrial television. Futhermore, the DBS system proposed for the 1990s will only have 3 channels, two of which will be NHK simultaneous broadcasts and the third will be controlled by a commercial group, Japan Satellite Broadcasting (JSB). This hardly presents an enormous addition to the nation's television menu though it does, as its supporters maintain, provide some expansion to the broadcasting system without disrupting the existing structures and endangering the livelihood of the commercial broadcasters.

Undoubtedly, there are some other strong arguments in favour of the DBS venture. Not only would it permit the development of an advanced industrial sector but it also impacts on the development of a large consumer industry linked to the provision of satellite dishes and related goods. As an industrial venture, it does have some enormous benefits.

But what would motivate a commercial consortium, such as JSB, to become involved in DBS? JSB is made up of some of Japan's leading hardware and financial interests, including the commercial TV networks and newspaper interests, and it clearly hopes to benefit from the popularity of DBS. If it does prove a success, DBS would become a major source of subscription and advertising revenue for the consortium. Indeed, subscription or pay-TV is the area of greatest interest for JSB and although there is as yet no indication of the nature of its future programme package, relying on subscriptions may push it to provide a popular service.[4]

The DBS venture is not, however, the only one intending to reach a new audience for television entertainment. Not unlike Europe, Japan also has organisations which intend to launch low or medium power communication satellites so as to offer a range of new services across the territory. Two satellite organisations, JCSAT and SCC, are currently in existence and both were licensed after the Japanese government deregulated telecommunications in 1985.

JCSAT is owned by C Itoh, MItsui and Hughes, whilst Space Communications Company (SCC) is an entirely Japanese venture controlled by Mitshubishi. Both organisations plan to launch their communications satellites in 1988 or 1989 using Ariane. Although intended as telecommunications satellites, they do aim to profit from the carriage of television traffic. They are reported as claiming that their main source of revenue in the early years will come from television traffic—including distribution of signals, news gathering transmissions etc.—before their data services begin to expand.[5] Unlike the DBS service, one will require a large TVRO of between 1 and 1.5 metres to receive the signal from these two satellites. It is worth noting that the use of satellites for feeding programmes to cable systems is by no means new to Japan. Satellite Video Services, a commercial satellite venture started in 1985, did just that though, according to Kobayashi, it was not particularly successful partly because of cable's lack of growth.

It is within this context that Japan's plans for DBS, and communications more generally, are formed and developed.[6] As the chapter below points out, there are overwhelming industrial and cultural reasons for exploiting DBS and these have helped frame the terms of reference of the DBS projects.

REFERENCES

1. Kobayashi, K. New Media in Japan Today, *Studies of Broadcasting*, NHK, 1985, pp. 7–29, p. 8.
2. See Tracey, M. in Negrine, R. (ed) *Cable Television and the Future of Broadcasting*, Croom Helm, London, 1985; Hills, J. *Deregulating Telecomms*, Frances Pinter, London, 1986. and Ito, M *et al, Broadcasting in Japan*, Routledge and Kegan Paul, London,1978.
3. Kobayashi, 1985, p.13.
4. Cable and Satellite Europe, London, October 1986.
5. Cable and Satellite Europe, London, October 1986.
6. Goto, K. 'Japanese Project for DBS Service', *Studies of Broadcasting*, NHK, March 1983, No.19, pp. 9–49.

Direct Broadcasting by Satellite in Japan: an Overview. By Yuko Nakamura

THE SIGNIFICANCE OF SATELLITE BROADCASTING

With a satellite in geostationary orbit about 36,000 km above the equator, it is possible to cover the entire length of the territory of Japan with one single beam. And, as the broadcast satellite's radio waves reach the earth at a large angle, one can overcome the obstacles of topography and buildings. Using the high frequency band available to satellite broadcasting has other advantages. For example, the wide bandwidth which is not ordinarily available through terrestrial broadcasting, offers the possibility of creating a high quality broadcasting system. Given these characteristics of broadcasting by satellite, one can note the following significant points concerning its development in Japan.

(a) An increase in the number of broadcast channels

The 1977 World Administrative Radio Conference (WARC-BS) allocated Japan eight frequency channels and the orbital position 110 degrees east longitude for the purposes of television broadcasting by satellite. It thus became possible to consider the development of eight new national channels of broadcasting so overcoming the limitations of terrestrial broadcasting. As the VHF and UHF frequency bands for terrestrial television broadcasting were already almost all used up, it had become increasingly difficult to develop or create new national broadcasting channels. Using a satellite for broadcasting significantly increases the available frequencies for such purposes.

(b) The possibility of a new broadcasting system

Broadcasting by satellite also makes it possible to run a high-definition television broadcasting service (HDTV). HDTV requires a wider frequency bandwidth than conventional television broadcasting and this is available via satellite systems. Other services which can be delivered by satellite include PCM sound broadcasting which exclusively uses one channel of television broadcasting and still picture broadcasting and both teletext and facsimile broadcasting. All these new services will be designed to meet the varied demands of the Japanese public.

(c) The speedy and economic introduction of national broadcasting

To run a national broadcasting service using a terrestrial transmission network, it is necessary to build a large number of relaying facilities. But with satellite broadcasting—once the satellite is launched with the necessary transmission and control facilities already installed on the ground—it becomes possible to create a high-quality national broadcasting system instantaneously and at low cost. However, one needs to bear in mind two difficulties that may arise, namely, the likely cost of purchasing and installing a satellite receiving dish which will have to be borne by the public and the need to provide broadcasting content which is as suitable for, and relevant to, the audience as terrestrial broadcasting is.

(d) Measures against disasters

In a country like Japan where disasters such as earthquakes and typhoons occur frequently, it is important that the transmitting facilities are able to withstand them and their effects. Satellite broadcasting systems, particularly those which employ several transmitting earth stations installed at numerous locations, can easily withstand the effects of disasters and so create a very reliable broadcasting network.

(e) The contribution to technological development and co-operation

In developing satellite broadcasting, many new and highly advanced techniques of research and development will have to be used. This will have a positive effect on the domestic advanced technology industry. Furthermore, if satellite broadcasting proves popular, this will encourage the development of a whole industry linked to the construction and marketing of receiving dishes. All these various industrial activities are likely to contribute to an enormous expansion in certain technological sectors as well as to lead to new forms of international co-operation in this field.

Figure 10.1: Evolution of Japan's satellite broadcasting system

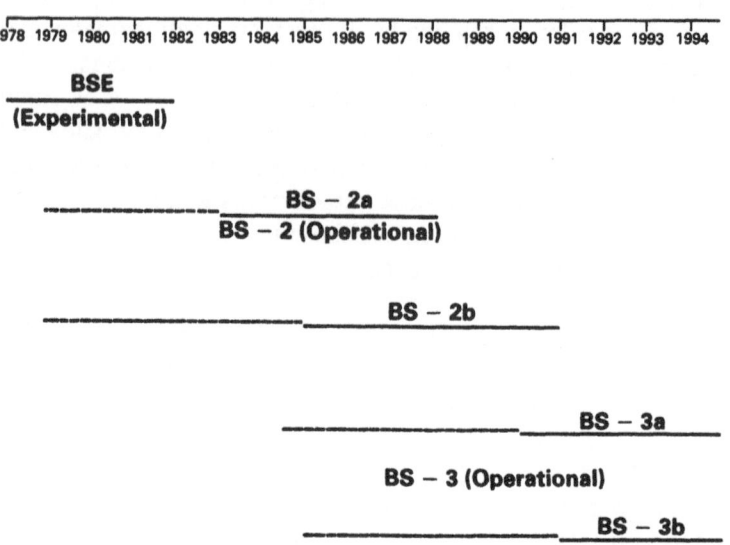

THE PRESENT STATE OF SATELLITE BROADCASTING IN JAPAN

Broadcasting by satellite in Japan has, to date, involved the following satellite systems:
—the experimental medium-scale broadcast satellite, BSE, which was launched in 1978, and,
—the BS-2 series, launched in 1984 and 1986, which were the first operational satellites to be used.
The next generation of broadcasting satellites, BS-3, will be launched in 1990 and 1991. Figure 10.1 sets out the evolution of these satellite systems.

The experimental medium-scale broadcast satellite, BSE

Studies relating to the development of the experimental broadcast satellite BSE began in about 1972. It was Japan's first experimental satellite and it had three main objectives. Firstly, BSE could be used to conduct tests on the use of satellites for the transmission of video and sound signals. Secondly, it would act as a preliminary to launching a large-scale satellite which would allow for the individual reception of television broadcasting; this would eliminate the existence of poor reception areas within Japan. Thirdly, the BSE would establish the sorts of operational techniques and facilities necessary for the launch and maintenance of a satellite broadcasting system before the operational BS-2 comes along. In the future, it could also be used for a 'broadcasting college'.

In 1973, the Space Development Project of Japan decided that the satellite should be launched by NASA's Delta rocket. The Japanese Ministry of Posts and Telecommunications (MPT) took the initiative for the project and, with the co-operation of other and related organisations, guided its development.

The National Space Development Agency of Japan (NASDA) took charge of the construction of the satellite while the Radio Research Laboratories of the MPT built the main earth station which was also used as an operational control centre. Finally, NHK was responsible for the construction of the transportable transmitting and receiving stations, the receive-only stations and the simplified receiving equipment.

The BSE was launched by the Delta rocket in April 1978. Following its launch, the MPT, NHK and NASDA all conducted a variety of experiments using the satellite. The performance of the satellite was as expected and it demonstrated that the satellite signal could be received satisfactorily using a small receiving antenna. The experiments also provided a great deal of useful information which was relevant to the introduction of the operational satellite.

BS-2

The BS-2 was Japan's first operational broadcast satellite. It was developed from about 1980 on the basis of the results of the experiments conducted on BSE. Its main objective was the elimination of the poor reception areas which existed within NHK's television broadcasting service. But BS-2 was also important for advancing the techniques relevant to the introduction and maintenance of satellite broadcasting systems generally and Japan's satellite broadcasting projects in particular. As regards its construction, BS-2 is almost identical to BSE in scale and it is the first broadcast satellite in the world designed for the individual reception of broadcast signals based on the WARC-BS 1977 frequency and orbital allocation plan.

The BS-2 system consists of two satellites, BS-2a and BS-2b. Both satellites were launched by N-II rockets from NASDA's Tanegashima Space Center.

BS-2a was launched on January 23rd, 1984. But due to a series of problems with two of the three transponders on the satellite, it was not possible to run the two channels of regular broadcasting which had originally been planned. Instead, a trial television broadcasting service was introduced in May 1984. This trial service used one channel and broadcast eighteen hours a day from 6 a.m. until midnight. This service mainly consisted of the simultaneous broadcasting of the terrestrial general television service although for a short period it did also carry programmes specially produced or arranged for the satellite service.

With the introduction of satellite broadcasting via BS-2a, it has now become possible to receive good quality television pictures in the outlying areas of Japan, such as the remote islands of Ogasawara, and in the mountainous areas of the Main Islands of Japan, as well as in other areas where it had previously been impossible to receive any television pictures. In those areas, even the trial broadcasting service had been welcome.

But the many features of satellite broadcasting make it important in other ways. For instance, the broadcast satellite is by nature accessible from any place in the country and this has been used to great advantage by the broadcasting organisations. On two occasions in recent years, a simplified transportable earth station was used for the on-the-spot relay broadcasting of important news events, namely, the earthquake which occurred in the western part of Nagano prefecture in August 1984 and the crash of the JAL jumbo jet which occurred in August 1985.

BS-2b was launched in February 1986 about six months later than the scheduled date. Because of certain difficulties with the main computer control system, it was necessary to use the spare one. As regards the transponders, these are operating stably and two channels of broadcasting have been available via BS-2b since December 1986. (BS-2a is now in a standby state.)

Up until July 1987, these two channels consisted mainly of the simultaneous broadcasting of the terrestrial television services—General Television and Educational Television. Since then these two channels have changed the content of their

Figure 10.2: Coverage of BS–2: Contour lines show diameter of receiving dish

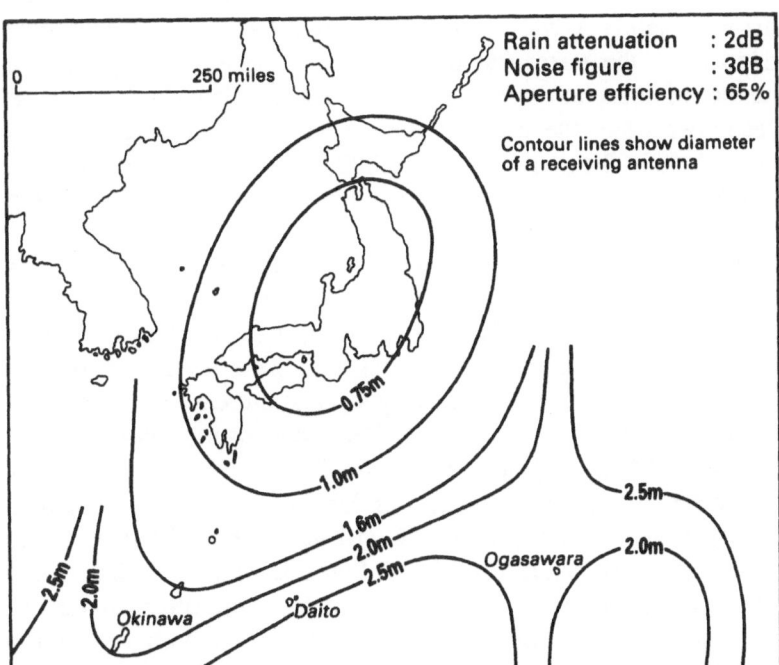

programmes. One channel is devoted to a mixture of programmes obtained from the General and Educational terrestrial services; the other channel is now a 24 hour channel devoted to programmes specially made or bought for BS-2b. These include world/international news fed by Intelsat from the USA, news fed from Super Channel and the BBC and business information, sports programmes and entertainment and music programmes. All these programmes have an international character and are proving very popular. As a result, it is expected that the numbers of viewers to satellite broadcasting services will increase rapidly.

The sizes of the satellite receiving dishes necessary for the reception of satellite television via BS-2 are shown in Fig.10.2. For most parts of the Main Islands of Japan, the required dish would be of about 75 cm in diameter; for the territory of Japan, but excluding the remote islands such as Ogasawara, the dish would be of about 1 metre in diameter.

It has been difficult to gauge the popularity of satellite broadcasting for two main reasons. Firstly, and until the launch of BS-2b, only one television channel was available via satellite and this was mainly a trial service. Secondly, that service was almost identical to the terrestrial broadcasting service. As a result, the number of households with satellite receiving dishes was small. At December 1986, the total number of satellite receiving facilities was 117,500. These consisted of 6,000 individual receiving dishes; 110,000 dishes for community reception and 1,500 for the reception of re-transmitted services on the islands of Daito and Ogasawara.

With the advent of BS-2b, and the availability of two television channels carrying some original and attractive programmes specially made for the satellite services, it is expected that these numbers will increase. NHK intends to make every effort possible to popularise satellite broadcasting so as to increase the number of households able to receive satellite broadcasting services to 1 million by 1990, the year in which BS-3 is to be launched.

But BS-2b has also presented an ideal opportunity to develop a variety of new techniques which will have applications in future broadcasting by satellite projects. Since its launch in late 1986, NHK has carried out numerous experiments using BS-2b and it will continue to do so in the future with special attention being paid to the following areas:

(a) High-definition Television (HDTV) broadcasting

One aim has been to obtain technical data so as to establish an HDTV transmission system using the MUSE signal reduction system. Another aim has been to increase awareness of, and to promote, HDTV through events such as Hi-Vision Fairs held in Tokyo in May 1987.

(b) Multi-channel PCM sound broadcasting

The main objective here is to obtain technical information which will facilitate the establishment of certain technical standards for permitting the carriage of 12–16 channels of stereo sound broadcasting on one television channel.

(c) Scrambling systems

In order to introduce some form of Pay Television, it is necessary to have some means by which the video and sound signals of standard television are scrambled. Such systems are now being considered by the Telecommunications Technology Council, an advisory body of the MPT, and experiments are now being carried out to obtain technical data so as to achieve these objectives.

In addition, it is likely that technical experiments on scrambling systems for HDTV broadcasting will be starting shortly.

(d) Data broadcasting

This is a form of data broadcasting which is multiplexed onto the data channel of the sound PCM signal. This is also being considered by the Telecommunications Technology Council. In the various experiments being conducted, technical information will be sought regarding this form of data broadcasting as well as the transmission of digital facsimile systems and programme identification codes.

BS-3

The launch of BS-3, the second generation of operational satellites, is planned for 1990, at about the time when the life span of BS-2 comes to an end.

The main objective of BS-3 is to take over the broadcasting services of BS-2 and to meet the increasing demands for more broadcasting and of a greater variety. The BS-3 series consists of two satellites, BS-3a and BS-3b. The launch of BS-3a is planned for the summer of 1990 and BS-3b will be launched a year later in the summer of 1991. Both will be launched by H-I rockets. At present, the design and the manufacture of BS-3 is in progress. Its major specifications are outlined in Table 10.1.

The uses to which BS-3 will be put were decided in November 1983. There will be three channels devoted to television broadcasting. NHK will use two of these channels in the same way that it is now using BS-2, and the third channel will be used by a newly established, and the first, private satellite broadcasting company. Originally, in 1983, fourteen companies had applied for the licence to run the private broadcasting satellite service but since then they have been integrated into one company and established as Japan Satellite Broadcasting Ltd. (JSB) in December 1984.

As with BS-2, the NASDA is responsible for the development and manufacture of the BS-3 satellite. The Nippon Electric Co. (NEC) is the main contractor and RCA/AE of the United States is a sub-contractor. The transponders and antennas on BS-3 are being manufactured by NEC and their design reflects the results of NHK's accumulated research over a long period in this field.

Table 10.1: Specifications of Japan's BS-2 and Bs-3

	BS-2	BS-3
Launch vehicle	N-11	H-1
Launch date	Jan. 1984/Feb. 1986	Summer 1990/91
Weight (in orbit)	about 350 kg.	about 550 kg.
Attitude control	Zero momentum three-axis stabilisation	Biased momentum three-axis stabilisation
Communication sub-system		
Frequency (Receive/send)	14/12 GHz	14/12 GHz
Number of Channels	2 (half redundant)	3 (full redundant)
TWT output power	more than 100W	more than 120W
Bandwidth	17 MHz	27 MHz
		1 (no redundant) more than 20W 60 MHz (Telecomm Experiment)
Life	5 years	7 years

SATELLITE BROADCASTING SYSTEMS: SOME TECHNICAL CONSIDERATIONS

The BS-2 satellite broadcasting system

The operational configuration of the BS-2 satellite broadcasting system is shown in Fig. 10.3.

(i) NASDA is responsible for the launch phase using its Tracking, Telemetry and Command (TT&C) equipment. In its operational phase, the Telecommunications Satellite Corporation of Japan (TSCJ) controls the position and attitude of the satellite from its control centre.

(ii) The two satellites (BS-2a and BS-2b) are positioned at 110°E longitude in geostationary orbit and illuminate most parts of the mainland of Japan with the maximum e.i.r.p. of 58 dBW by using a shaped beam antenna with three radiation horns.

(iii) The main earth station at the NHK broadcasting centre in Tokyo is the key facility for transmitting television programmes to

Figure 10.3: Japan's BS–2 satellite broadcasting system

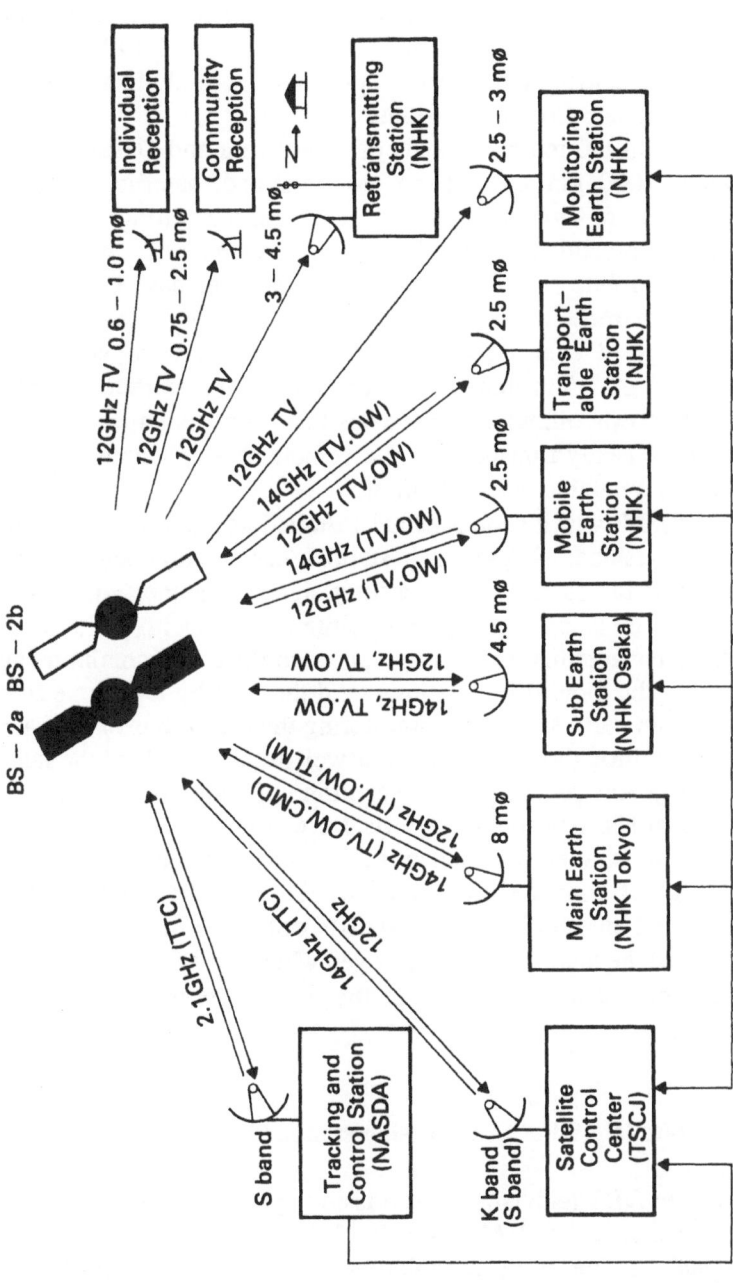

the satellite. The earth sub-station at the NHK Osaka is used as a backup to the earth main station. The earth main station is equipped with two antennas of 8 m and 5 m in diameter, full redundant 14 GHz transmitters of maximum output power of 1.4 kW as well as various control and supervisory facilities including Telemetry, Tracking and Control facilities.

(iv) In the remote islands of Ogasawara and Daito, which are about 1000 km away from the mainland, satellite programme services are available through VHF retransmitting stations and satellite receiving dishes of between 3 to 4.5 metres in diameter.

(v) Mobile and transportable earth stations have been prepared for the purpose of live broadcasting in times of disasters or for the coverage of events for local programmes from anywhere in the country. A 2.5 m diameter antenna and a transmitter of 200W or 500W power output have been designed to be mounted on a vehicle or to be easily transported by a vehicle or a helicopter. These are readily available when required.

(vi) Monitoring stations are installed at 5 points, including the remote islands. These earth stations can monitor and record the level of the satellite broadcasting signals together with levels of rainfall by using a precision satellite receiver with a 2.5 m to 3 m diameter antenna as well as data recording and communications units. The monitoring stations send the data by telephone lines to the main earth station for monitoring the conditions of the satellite service throughout the country as well as the attitude of the satellite.

(vii) As regards the reception of signals by individual households, determined efforts have been made to improve the aperture efficiency of antennas (65% to 75%) and the NF (noise figure) of converters (3 to 2 dB). High quality pictures can be obtained by satellite receiving dishes of 75cm in diameter in most parts of the mainland of Japan even when taking rain attenuation of 2dB into account (for 99% of the time in the worst month of the year).

Satellite television transmission standard

All the DBS television transmission systems should comply with the decisions of WARC-BS 1977. For the transmission of video signals, a frequency modulation with the 525 line NTSC video signal, which is extended to 4.5 MHz, was adopted in 1983, as shown in Table 10.2. As regards sound transmission, the QPSK

Table 10.2: Satellite television transmission system

(a) Television picture

Television system	525-line/NTSC system
Video frequency bandwidth	4.5 MHz
Type of modulation	FM
Frequency deviation of carrier	17 MHz p-p
Pre-emphasis	CCIR Rec. 405
Frequency deviation of carrier by energy dispersal signal, symmetrically triangular, with frequency of 15 Hz	600 kHz p-p
RF bandwidth	27 MHz

(b) Television sound and data

Parameter	Mode A	Mode B
Coding (PCM)		
Sound signal bandwidth	15 kHz	20 kHz
Sampling frequency	32 kHz	48 kHz
Quantising and companding	14/10 bits, near-instantaneous companding (5 ranges)	16 bits linear with range code
Sound emphasis	50/15 usec	
Multiplexing		
Bit rate	2.048 Mbits/s	
Number of sound channels	4	2
Additional data capacity	480 kbits/s	240 kbits/s
Modulation		
Modulation method of subcarrier	4Φ-DPSK	
Subcarrier frequency	5.727272 MHz	
Frequency deviation of main carrier by the subcarrier	± 3.25 Mhz	

(Quadrature Phase Shift Keying) system was adopted for high quality. This system provides for the sound sub-carrier to be modulated by the sound PCM signal. Two kinds of sound codings can be selected: Mode A for the transmission of 4 channels (15 kHz bandwidth each) or mode B for high quality transmission of 2 channels (20 kHz bandwidth each). Additional data channels are accompanied with sound channels such as teletext broadcasting and facsimile broadcasting. Basic transmission parameters for television sound and data are shown in Table 10.2.

SATELLITE BROADCASTING: SOME FUTURE POSSIBILITIES

Satellite broadcasting at the end of the 20th century

For satellite broadcasting to become a popular and successful medium in the future, it will be necessary for the following developments to take place:
—the broadcasting service and its programme content will have to be versatile and varied so as to meet the needs of the audience;
—high definition television (HDTV) will have to be introduced to attract the public;
—small receiving dishes will have to be introduced. For this development to take place, it will be essential to have a high output power satellite as well as a further improvement in the already high performance of receiving dishes.
—the cost of these small receiving dishes will have to be much lower than at present;
—there will have to be an improvement in the reliability of satellite broadcasting systems as well as a development of such systems at a much lower cost.
    In addition to the above, satellite broadcasting systems will need to respond to the demands of the country for a greater range of broadcasting content. This will be achieved only if there is an increase in the number of useable channels for broadcast television and an exploitation of those channels in such a way as to make the most of the unique characteristics of satellite broadcasting. The emphasis must be, therefore, on an expansion of channels and their use in new and varied ways. The implementation of the above measures is essential if direct broadcasting by satellite is to

succeed. If these measures are not implemented prior to the launch of commercial communication satellites in 1988, then it will face severe competition. These commercial communication satellites are intended for the transmission of television programmes to cable television systems and so are likely to have an enormous impact on the popularity of municipal community antenna television systems. Such systems are considered to be the most important means by which to provide more television channels, channels far in excess of those available via direct broadcasting by satellite.

The key to the future success of satellite broadcasting—and an important means by which to increase its popularity—is the development of HDTV. With terrestrial broadcasting, it is very difficult to introduce HDTV for economic and technical reasons. But an important feature of satellite broadcasting is that it provides a wide bandwidth which can be used for HDTV. Nevertheless, there are some crucial considerations to take into account before HDTV can be introduced and these have a bearing on its future availability and popularity. These considerations include the frequency band to be used by future satellite broadcast systems, the development of low cost satellite receiving dishes and high definition television studio and transmission standards.

With regards to the frequency band which the satellite system can use for HDTV broadcasting, there are three possibilities: (i) using the 12 GHz band as currently planned, (ii) using the 22 GHz band, or, (iii) at some future date, making some sort of change to the current plan to use the 12 GHz band.

At present, NHK is considering adopting the first of these three possibilities, namely, using the 12GHz band for HDTV. It is hoped, therefore, that the HDTV service will start with the first stage of the BS-3 satellite system and that it will use the baseband signal reduction technique called MUSE which has been developed by NHK. As a result of this proposal for an early start to HDTV, the Broadcast Technology Association (BTA), is looking into the establishment of provisional production and transmission standards. The Telecommunications Technology Council of the MPT will follow the BTA recommendations in setting the future standards for HDTV.

As for reducing the cost of the receiving dish — a critical factor for the mass take-up of the satellite broadcast service — it will be important to increase the strength of the signal received on the earth by increasing the power output of the broadcast satellite. This will have the effect of reducing the size of the dish required for satellite

265

television reception. It would also make it much easier to install the (smaller) dishes.

The technical features of the satellite receiving dishes,namely, the efficiency of the aperture and the noise-figure (NF), are, in fact, continually being improved. With the existing satellites of 100 W, it is already possible to obtain good television pictures across the main territory of Japan—but excluding the remote islands—by using a parabolic dish of about 50–60 cm in diameter. Such a power output is considered to be adequate for broadcast satellites up to the end of this century.

In order to ensure a continuity in satellite broadcast services in the future, it will be necessary to develop and to plan the launch of the next generation of satellites, that is, post BS-3, before the life span of BS-3 comes to an end. At that point in the future, it should be possible to use a rocket (the H-2) which would be capable of launching a larger class of satellites of 2 tons. These satellites could have a whole range of different features depending on the number of transponders required, their power output, the weight of the satellite and the length of its working life.

A typical, next generation, large broadcast satellite would probably fall into one of the following three groups, each one of which has unique characteristics.

(i) A one ton satellite, offering 3–4 channels of broadcast television and with a working life of seven years.

This system would provide for 3–4 200 W transponders (with an additional 3–4 transponders as spares) in the 12 GHz band. This would give national coverage. It could also offer two 20 W transponders (with two spare) in the 12.5 GHz band and one 60–200 W transponder (spot beam) in the 22 GHz band.

(ii) A 1.5 ton satellite offering 6 channels of broadcast television and with a working life of ten years.

A satellite of this size would offer six 200 W transponders (with an additional six as spares) in the 12 GHz. Such a service would give national coverage. This satellite could also provide services in the 12.5 GHz and 22 GHz bands of the same type as available under the above satellite.

(iii) A two ton satellite, offering 8 channels of television broadcasting, with a working life of ten years.

The largest satellite would provide for 8 200 W transponders (with an additional 8 as spares) in the 12 GHz band, two 20 W transponders (with 2 spares) in the 12.5 GHz band and 1 to 4 60–200 W transponders (spot beams) in the 22 GHz band.

Satellite broadcasting in the 21st century

As we enter the 21st century, reusable launch vehicles and space stations will find increasing use and the satellite systems which will be provided in the future will change fundamentally from those in current use. It may be that the broadcast satellite of the future will become a type of economic geostationary platform and serve a variety of frequency bands. The stationary platform will be a large structure, possibly constructed in space, and it will allow for new developments by simply mounting a large antenna onto the platform itself. Finally, the platform will have all the facilities necessary for repairing faulty equipment as well as supplying fuel.

Of all the frequency bands available for satellite broadcasting, the most interesting one, and by far the most powerful one, is the 22 GHz band. This band does, however, have a problem in that the signal attenuation due to rainfall is larger than the signal attenuation in the 12 GHz band. Nevertheless, this band is likely to be used because it does offer many advantages such as increased versatility for satellite broadcasting in the future. In addition, as the satellite's spot beams can be focused easily, it will be suitable for sub-national (satellite) broadcasting by dividing the whole country into several different areas. In addition to HDTV, this frequency band will also be suitable for Integrated Services Digital Broadcasting (ISDB) in which all kinds of broadcasting services are integrated and digitally transmitted. This would produce the most advanced form of digitalisation of broadcasting to date. In fact, the proposal for ISDB was put forward by Japan to the CCIR and it is currently being investigated by that group with a view to its international implementation.

CONCLUSION

Since the first operational satellite broadcasting system was launched in January 1984, Japan has been able to acquire a great deal of technical and other information about this form of broadcasting. The satellite is currently—and successfully—continuously transmitting two high quality television services, accompanied with PCM sound, and available throughout the country. The size of the audience able to receive these services is increasing steadily.

But at the same time, a great deal of effort has gone into a series of technical experiments in order to develop other advanced uses of satellite broadcasting. There have been numerous experiments and demonstrations of such things as HDTV, multi-channel PCM sound broadcasting, facsimile broadcasting and more advanced forms of data broadcasting. All these have been tried out during the vacant hours of the BS-2 satellite, that is, when it is not broadcasting the television services.

So, despite the risky and rather costly nature of satellite broadcasting systems, it has many advantages and offers so many exciting and attractive services that it becomes a worthwhile project.

The next stage of development is BS-3. The design and manufacture of this next generation of satellites is in progress and it is scheduled to be launched in 1990. Three channels of broadcasting will be used by two organisations, NHK and the newly formed Japan Satellite Broadcasting company. All concerned are making every effort possible to ensure that this new satellite project will be a successful one and that HDTV, one of the most attractive features of satellite broadcasting, will get off to a good start.

Index